AMERICAN DUNKIRK

James Kendra and
Tricia Wachtendorf

AMERICAN DUNKIRK

The Waterborne Evacuation of Manhattan on 9/11

TEMPLE UNIVERSITY PRESS
Philadelphia • Rome • Tokyo

TEMPLE UNIVERSITY PRESS
Philadelphia, Pennsylvania 19122
tupress.temple.edu

Published 2016

Library of Congress Cataloging-in-Publication Data

Names: Kendra, James M., 1965– author. | Wachtendorf, Tricia, 1972–
Title: American Dunkirk : the waterborne evacuation of Manhattan on 9/11 / James M. Kendra and Tricia Wachtendorf.
Description: Philadelphia : Temple University Press, [2016] | Includes bibliographical references and index.
Identifiers: LCCN 2015039525 | ISBN 9781439908204 (cloth : alk. paper) | ISBN 9781439908211 (paper : alk. paper) | ISBN 9781439908228 (e-book)
Subjects: LCSH: Disaster relief—New York (State)—New York. | Emergency management—New York (State)—New York. | September 11 Terrorist Attacks, 2001.
Classification: LCC HV555.U62 N5233 2016 | DDC 974.7/1044—dc23
LC record available at http://lccn.loc.gov/2015039525

♾ The paper used in this publication meets the requirements of the American National Standard for Information Sciences—Permanence of Paper for Printed Library Materials, ANSI Z39.48-1992

Printed in the United States of America

121819P

Contents

Prologue

Manhattan is an island. This geographic detail seems less important than it once did, as the rough wooden waterfronts shown in crinkling photographs have given way to glamorous financial-services offices, high-end apartments, and stylish shops and restaurants. The waterfront, especially around Lower Manhattan, has become a fun place for tourists, joggers, business-lunchers, and hand-holding couples. But hundreds of boats and barges continue to ply the waters around Manhattan. Tugs, the famous Circle Line sightseeing boats, dinner cruise boats, ferries, and other craft of every description form the background scenery for people standing on the shoreline looking out across New York Harbor. Beyond the fun and urban excitement is the workaday maritime world of diesel oil, coveralls, steel-toed shoes, and dirty gloves. The shift to commerce grounded in digital, virtual, and financial ephemera has not lessened the need for food and fuel around the city. These boats are not just decorative accessories placed on the waters to give tourists something to look at; they link the city to the regional, national, and global economies.

New York City's maritime setting came into sharp relief during the terrorist attacks of September 11, 2001. Anyone who watched television, read a paper, or fought through packed bandwidth to try to get on the Internet knows what happened. Early reports told of a small plane striking the North Tower (World Trade Center 1) at 8:46 A.M. In fact, as we would learn, this was not a small sightseeing plane but rather American Airlines Flight 11—bound from Logan International Airport in Boston to Los Angeles. At 9:03 A.M., when United Airlines Flight 175—also en route from Boston to Los Angeles—struck the South Tower (World Trade Center 2), the hostile nature of the event became widely apparent. Over the course of the next hour, two more airplanes were hijacked: American Airlines Flight 77 en route from Dulles International Airport in Virginia to Los Angeles struck the Pentagon at 9:37 A.M.; and United Airlines Flight 93 en route from Newark International Airport to San Francisco crashed into a field in Shanksville, Pennsylvania, when passengers tried to overpower the hijackers. In all, according to the 9/11 Commission Report, a total of 2,973 were killed, excluding the hijackers (National Commission on Terrorist Attacks upon the United States 2004).

That morning, every eye and every camera in the world, it seemed, was focused on the glowing towers and then, in turn, on their unbelievable collapses—Tower 2 at 9:59 A.M. and then Tower 1 at 10:28 A.M. And because most of us were watching the burning and disintegrating towers, and then the astounding, shimmering cloud of dust, we missed something remarkable happening nearby, along the shoreline of Manhattan, from Chelsea Docks on the West Side Highway around to the Staten Island Ferry terminal and up the eastern waterfront.

Some people were not just watching the towers or watching the skies for more planes. Mariners' eyes move constantly, taking in the water, other boats, and the shoreline. And in that gaze, they saw a need they could help meet. Across New York Harbor, different ideas flashed like sparks, which ignited a collective understanding when boat operators and waterfront workers realized that they had the skills, the equipment, and the opportunity to take definite, immediate action in responding to the most significant destructive event in the United States in decades. Their spontaneous convergence toward the downtown area succeeded in moving hundreds of thousands of evacuees from around the southern reaches of the island.

On its own, this would be an amazing story—a Dunkirk-style evacuation in an improvised fleet, without any significant accidents. Indeed, several people whom we interviewed for this study specifically referred to the event as "like Dunkirk," invoking reference to the successful 1940 evacuation of several hundred thousand soldiers from France over several days during World War II. But the remarkable story continued on the other side of the Hudson. Once on the New Jersey side, evacuees had nowhere to go. Some people still in New York City were hesitant to get into boats, fearing they would be stranded. One of the tour boat operators got his business partner—a bus company—to pick up evacuees and take them to transit connections; soon, other bus agencies began mobilizing independently to help in similar ways. It did not take long for boats to begin to return to Manhattan with emergency workers and supplies, and, as the evacuation wound down, some boats stayed in service to shuttle anything the response workers needed. Others helped out however they could while docked nearby. The dinner boat *Spirit of New York* offered its facilities to weary rescue workers, and the *John J. Harvey*, a retired fireboat, used its vast pumping capacity to support firefighting efforts. An ad hoc transportation system evolved, carrying passengers and equipment and supporting the "official" rescue efforts.

None of these efforts was planned. Instead, the waterborne operations were improvised and spontaneous: the emergent invention of the necessity of the moment, as people taken by surprise pulled together, defined the disaster in their own terms, and determined a way to help. It is far too trite to say that lessons can be learned from this event: Every disaster offers lessons. But understanding the evacuation—and all the elements of people and organizations that made it possible—has larger implications for the entire practice of disaster management in the United States and beyond.

In this book, we examine how maritime workers identified what they had to do and how their interpretation of what was needed helped connect them to others who were working with their own understanding of what was happening on the morning of September 11. Most of them had never been involved in any water evacuation from Manhattan or anywhere else, although a few had been involved in the much smaller boat lift of people after the 1993 bombing of the

Twin Towers. None was aware of any plan in place for such an event, yet they simultaneously understood that their skills were relevant.

Our goal is to examine how people, as individuals, groups, and formal organizations, pull together to respond and recover from startling, destructive events. When foresight, planning, and practice are defeated, what do people do next? As we show, the participants in the boat evacuation reinterpreted their new surroundings and started acting, relying on tacit knowledge and latent resources to adjust to a damaged landscape. They were "regular" people, average citizens, or members of what may formally be called civil society. What can they teach us about not only surviving but also thriving in the face of calamity?

In many disasters, events that seem larger than life are dealt with and managed, mostly, if not always most visibly, by the actions of ordinary people, extending their knowledge, skills, and resources to address small elements of a big, perhaps catastrophic, problem. In some cases, these are people charged with official responsibilities who go beyond what they had expected they would ever need to do. Others are people who never considered themselves disaster responders yet step up to the plate. The people we include as examples in this book epitomize stories we heard again and again, told a little differently on occasion, from a different part of the harbor or with a different color of narrative, but with remarkable similarity nonetheless. We discuss the conditions that allowed and even fostered the evacuation to unfold the way it did and explore the ways that people who were physically separated from one another began to share a sense of what was needed. We show, in particular, that handling a disaster is not something that happens apart from the community; it involves everyone. Our vulnerability to disaster is threaded through the social and physical systems of our towns and cities; our resilience and recovery potential is found in those same systems, animated by the enthusiasm of private citizens to cross over into activities that are generally considered "official."

We heard repeatedly in our interviews such comments as "We just did what needed to be done" or "We did what we had to do," not only from mariners but also from the people with whom we discussed their stories. A few weeks after the attacks, for example, we encountered a group of bicycle couriers who wanted to help at Ground Zero. They were rebuffed by officials—perhaps rightfully so, given their skill sets and the hazardous environment at the pile—so instead they delivered sandwiches and coffee around the secured perimeter of Lower Manhattan. They took what they had—a resource (a bike), a skill (quickly navigating the city by bike), a network (an organization of street-wise cyclists)—and identified how it might be useful, even if just for a short time. As one of the mariners we talked with said, if you had local knowledge, at least at the beginning, the opportunity to find a way to help was available.

Such actions raise obvious potential objections and concerns regarding risk, liability, and security. Could a person sneaking in be a threat? Of course. In some settings, such as Israel, there is a strong danger of secondary attacks after a bombing, specifically targeting response personnel. Still, the overwhelming history of postdisaster helpers shows that security incidents have not occurred after disasters, even in settings where they might be expected. It is impossible to estimate the number of volunteers who came to New York City, but they came from all over the country, and from around the world. Many of them were first responders—fire, police, emergency medical technicians (EMTs)—who responded to the call of professional kinship. They did not necessarily know the city, but at least they had expertise. Professional emergency managers worry more that lay helpers will get in the way or that their well-intentioned but uncoordinated efforts will cause a breakdown in control, waste scarce resources, or crowd the areas so that qualified responders cannot gain access. But in the maritime operations on 9/11, very little evidence suggests that such problems occurred. In desperate situations, there may be no option but to let volunteers do their work. And indeed, in many cases, the work of spontaneous volunteers bringing particular skills, knowledge, and resources enables the formal responders to succeed.

Our many conversations with those who helped that sunny, mild, tragic, inspiring day led us to conclude that managing a disaster means tackling an enormous event in bite-sized chunks. The mariners we spoke with were stunned by the enormity of the crisis and the grim and sorrowful novelty of what they were seeing. But although the attacks brought unprecedented destruction to New York City, in their actions, waterfront workers and others throughout the city did not approach it at that scale. How could they? Regular people, by definition, are regular. Instead, they brought their usual knowledge and skills to the disaster, shattering the enormity of the disaster itself into smaller pieces that they could pick up. It was just that *a lot* of them did it, all over the city.

All disasters are cases of the ordinary achieving the extraordinary—in the case of the boat evacuations, with tremendous success. The principal insight that we hope readers take away from this book is that people have more capacity than they think. People can do a good job helping out in disasters just by being themselves, by finding or making a chance to help. It suggests that what we need as a society is resilience grounded in diverse institutions, industries, and skills, a resilience that is energized by a willingness to assemble those fundamentals into new systems. While a disaster appears to be something best managed by officials with technical expertise, the fact is that most disaster-related tasks are everyday tasks. Lifting boxes, loading trucks, answering phones, delivering coffee, sharing food and blankets, sweeping up broken glass—these, too, are disaster jobs that anyone can do. And because a disaster by definition affects an entire community, practically everyone has something he or she can do that could be helpful.

Particularly since Hurricane Katrina, the popular press has questioned the competence of local populations to manage themselves in disaster. Supposed postdisaster chaos, violence, or social collapse generates calls for more immediate military involvement in a disaster area. While undoubtedly there are legitimate roles for law enforcement and the valuable personnel and materiel of military units, assuming that only those units are suitable for the postdisaster environment is likely to undercut the quick application of local resources. Indeed, the success of the waterborne evacuation of Manhattan on September 11 in part depended on the willingness of the U.S. Coast Guard, harbor pilots, and harbor police to allow for a decentralized response. Even with an eye for security and safety, they were still able to recognize the value of an improvised citizen response to the terrorist attack.

Planning is always essential, and no one would argue that formal emergency management agencies do not have a mandate to think ahead. At the same time, however, as a nation we have moved toward ever more tightly controlled disaster environments in which the roles of any and all responders are specified in advance and volunteers are registered ahead of time. We seem to have forgotten that improvisation is an important element of any disaster response. The success of our formal responses often depends on latent capacities in our own communities. In the response to 9/11, we saw large-scale, decentralized, emergent, grassroots efforts across the New York metropolitan region to help survivors, to treat the injured, to get people home, to put people in contact with their worried families and friends, to find them a place to stay if their homes were damaged or full of dust or without utilities. The waterborne evacuation of September 11 contradicts any doubts about what people can achieve in a disaster.

The cliché is well known and often told: September 11, 2001, began like any other day, and that was as true for us as disaster researchers as it was for so many others. Neither of us was in New York on that crisp, clear, surreal day. Jim was on his way to work at the Disaster Research Center (DRC) at the University of Delaware in Newark, Delaware, when the first news came over the radio that a small plane had struck the World Trade Center. Tricia walked the half block down the street from her apartment to the office. By the time each of us arrived, the first reports of the North Tower being struck had begun to emerge in the news. Each of us thought back to earlier that summer when we had individually made the long elevator rides up the towers: Jim up World Trade Center 1, or the North Tower, and Tricia up World Trade Center 2, or the South Tower. Jim had visited the city's new and state-of-the-art Emergency Operations Center located in World Trade Center 7, just across the street from the Twin Towers. Afterward, he had joined colleagues for dinner at Windows on the World, very much enjoying an expensive bleu-cheese burger and not particularly enjoying his first—and to this day only—martini. It had been a successful visit, with interesting discussions with emergency officials at the top of their game. Tricia, after many visits to New York City, had finally relented to make the touristy trip up to the Top of the World Observation Deck with her teenage nephew visiting from the Canadian prairies. She had smiled with nostalgia for her own first visit to New York years earlier, when she had looked around in amazement at the expanse of skyscrapers, people, and buzz of city life that is Manhattan. And she had conceded her own awe as she looked out from above at the incredible view of a city she loved.

As Jim parked his car and headed into DRC, and as Tricia headed for a computer, the second plane struck. Suddenly what had been a strange accident became something exponentially larger. Like the rest of the world, we were still grappling with *how*. But as researchers working at a center with decades of experience studying the early phases of disaster, this question also led us to ask *what* we as researchers needed to focus on. Jim ran across the parking lot. The scene at DRC must have been just one example of millions of similar scenes: phone calls, people trying to put together the pieces of news they had heard from various sources. In what seems an obvious oversight in retrospect, DRC did not have a television set, so everyone went across the street to watch the news at the University of Delaware's media center.

As disaster researchers, our task—at least at first—was to observe what was happening. Since this kind of fieldwork begins with gathering preliminary information from the media, we took notes based on television coverage to determine who seemed to be involved. Who were the main decision makers whom we would later attempt to interview? Where would we find documents or logbooks that tracked shifts in information as it became available to public officials? What would be some principal locations to visit once we made our way into the field? We did not know what we might find, of course. The key was to keep an open mind, to be alert for the features of the unfolding events that would best help expand our understanding of disaster management.

And then we sat awestruck. What was already a disaster suddenly expanded into an urban calamity. We cannot really say that it was unparalleled, since cities are regularly devastated by natural or human forces. But the gruesome novelty of the attacks, the surprising completeness of the towers' collapses, and their sequential pummeling of the surrounding neighborhood brought a new dimension to modern disaster. And all of it was unfolding live in front of an international television audience.

We kept on taking notes. Before graduate school, Jim had been a merchant marine officer, and his service aboard ship made noting the time of events an automatic task. In looking at his notes, you can see without even reading them when the first tower collapsed, signaled by the suddenly hasty penmanship as he scribbled what he saw. Tricia, who was DRC field director at the time, switched into "work mode." Perhaps that came naturally with a stoic farm upbringing, but there was some comfort in being able to fall back into the research steps and tasks that would be necessary to quickly deploy a team.

Stepping outside to walk back to the office, the silence of the town was unsettling, too, in its own way. By then all the nation's aircraft had been grounded, so there was none of the usual noise from the skies. Even the street traffic was quiet, as though people were driving on their tiptoes, sneaking their cars along the roads. Shocked as we were by what we had seen, even though it was on TV, training and well-learned procedures for disaster fieldwork provided the starting points. DRC's director at the time, Kathleen Tierney, had attended a meeting in Buffalo the day before and was stranded there after her plane, like all other aircraft in and entering U.S. airspace, was grounded. She was trying to contact our funding agencies while simultaneously trying to find a car rental location that still had a vehicle so she could drive home to Delaware—a story that was repeating itself around the United States. A 2:00 P.M. conference call with Kathleen, Jim, Tricia, faculty members Joanne Nigg and Benigno Aguirre, and DRC founders Russell Dynes and Enrico L. Quarantelli confirmed that DRC would send a team. Jim and Tricia kept trying their own contacts as well as those suggested by others. They mobilized the graduate and undergraduate students to canvass the media for names, addresses, phone numbers, and locations of emergency-response activities and also to scout out how to get access to New York despite the closure of many roads, bridges, and rail lines. We already had a project underway on studying resilience in communities—what makes a community less likely to have a disaster or more likely to recover from one? What, then, would we find in New York when we got there?

At this point, of course, we did not know exactly what we were looking for. We knew the basic parameters of community resilience, but we did not know how these would play out in the specific situation that was unfolding in the city. The first stage of disaster fieldwork is just basic fact-finding. Conditions on the ground, interpreted through the education and interests of the scholars, shape the subject and scope of the ongoing research. But we needed to get to New York because it is important, in disaster fieldwork, to get to the disaster area quickly. As much as possible, we want to see emergency operations in real time, to know the context of challenges confronted and decisions made. We were trying to get in touch with the emergency officials we knew, by phone and e-mail, but not surprisingly there was no response. Although we did not it know it at the time, New York's Emergency Operations Center had been abandoned early in the crisis, so no one was there to get our calls or see our messages, even if their communications lines had still been up. After a flurry of phone calls, a team of five DRC researchers arrived in New York City two days later. Jim and Tricia stayed on for a week and then made repeated trips over much of the next two months. Thanks to the incredible generosity of time and spirit of our emergency management contacts who had miraculously survived the attacks despite their proximity to them, we were given tremendous access to response meetings, operations centers, staging areas, and even Ground Zero.

A year later, we returned to conduct in-depth interviews with more than sixty key responders and decision makers, learning more about the many examples of improvisation that bolstered the effectiveness of the response and that had impressed us during our fieldwork. During the quick response research and later, during the interviews, we heard accounts about the boat evacuation and resolved to follow up on it if we could. Our in-depth research on the boat evacuation began in 2005.

Since that time, we have interviewed one hundred people with direct or indirect involvement with the waterborne evacuation. Most of these people were mariners, waterfront workers, harbor pilots, and Coast Guard officials. Others were ferry company office staff, emergency response workers, or people who had worked with the bus operators who eventually helped get evacuees closer to home. We focused on the helpers rather than the evacuees themselves, in part because evacuees were difficult to locate as time passed and in part because of the focus of our study. It is difficult to know how many helpers participated in the boat evacuation or subsequent maritime responses or their level of involvement (over nine hundred people were awarded medals or ribbons by the U.S. Department of Transportation for their involvement, including participants from some fifty organizations). It is even more difficult to count how many people participated in the shore-side operations that complemented the mariners' efforts. Still, we talked with a large and diverse group of participants who provided a rich understanding of the event.

At times, we conducted interviews by telephone; years after the disaster, some of the key participants had moved on to other states. But most of our interviews took place on their boats, in offices, in restaurants, and on the waterfront. As part of the interviewing process, to refresh the participant's memory and to provide a source of reference, each interview was conducted with a nautical chart of the harbor spread out in front of us. There, we could look at geographic features and indicate the movements of people and boats throughout the evacuation and boat lift, gently encouraging our maritime responders to mark up the charts with a Sharpie despite their inherent aversion to writing in ink even on photocopies of these important navigation sources.

We visited places where evacuees embarked or were disembarked, sometimes with those we interviewed (again, as a way to refresh their memories). We rode a number of the ferries and water taxis, timing the passage and the time it took to load and unload passengers (for example, it took *The Little Lady* about fifteen minutes to make the round trip to Liberty Landing) to feel confident in the number of evacuees that had been cited in various sources. We reviewed hundreds of photographs, newspaper articles, news accounts, e-mails, and segments of videotape, most of which were generated on the day of the attacks or within the weeks that followed. We augmented our interviews with eighteen conducted by David Tarnow, a historian who compiled an oral history for the South Street Seaport Museum two months after the attacks. In this work, we have developed the most comprehensive documentation of the evacuation available, and we have been careful to use these varied sources of information to triangulate or check the information we heard in our interviews, just as we compared the information we heard across interviews. Many of the mariners were excellent narrators of their involvement. The quotations we use in this book are drawn from our interviews with these participants unless otherwise noted.

We would be remiss if we failed to note the tremendous generosity of these participants. Their willingness to share their stories and experiences from that memorable day made this project possible. These were individuals who stepped up to help—like so many others on 9/11—and in this project, they recognized a potential opportunity to help again. But we also believe that they appreciated a story less told. Considerable attention has been directed at the commendable efforts in the Trade Center Towers, at Ground Zero, in the planes, and at the Pentagon in Arlington, Virginia, the site of one of the other attacks. The story of the boat evacuation and subsequent supply lift is a lesser-known story, so the participants were happy to share it. For this, we are thankful.*

* Missing from this work, notably, are the stories of those in the harbor community who did not participate in the evacuation. Anecdotally, we heard of boat workers who opted to return home that morning rather than join those who were deploying to evacuate others or those who chose to evacuate their families by boat and remain with them rather than leaving them. We strongly believe that their responses should not be diminished alongside their counterparts who did respond by participating in the evacuation. We cannot be certain we would not do the same, opting to return to a spouse at home or evacuate and remain with two young children as the devastating events unfolded. It would be easy for someone to conclude that his or her boat was too small or that the waterfront was already too crowded for his or her efforts to be of any significance. We attempted to interview several maritime workers who were in the vicinity and did not participate in the evacuation but were unsuccessful in securing their participation. This work would have benefited from their stories and a greater understanding of how people opt out or self-organize for very different ends. Our purpose here is not to mythologize the evacuation and boat lift at the expense of those who chose a different path but rather to understand how such a response became possible alongside other paths.

1 / Making Sense of Disaster

On September 11, 2001, an estimated three hundred thousand to five hundred thousand commuters and residents evacuated from Lower Manhattan in an ad hoc flotilla of ferries, tugs, workboats, dinner cruise boats, and other assorted harbor craft. Many of the evacuees were Manhattan commuters attempting to return home, while others were residents or employees of Lower Manhattan trying to flee the hazardous conditions that had quickly developed in this part of the city. In the confusion surrounding the attacks, people left the towers and the general Lower Manhattan area in all directions. While some hiked north, others trundled eastward across Manhattan and then across the Brooklyn Bridge. But those who walked or ran south and west were brought up short by the waterfront. Soon, members of the city's maritime community saw the growing need for transportation and realized they could help.

The boat lift began even before the collapse of the Twin Towers. Some ferry captains arriving in Manhattan saw or heard about the unfolding emergency and turned around with their passengers rather than disembark them, while others picked up passengers who reached the docks early in the growing crisis. A few mariners reported asking the Coast Guard for authorization at some point to approach the waterfront and pick up evacuees, but others did not. Some, like Captain Rich Naruszewicz, were in the vicinity when the first plane struck. Naruszewicz's boat was heading with passengers to Thirty-Fourth Street. He kept the passengers on board, swung back around to Pier 11, and picked people up for the trip back to Highlands, New Jersey, some twenty miles away. As the morning went on, decontamination squads began meeting the passengers on the Jersey side to help clean off the ashes and dust. Sometimes a clean change of clothes was available. After that, passengers could receive first aid, water, and counseling. The boats, meanwhile, were refueled, and then back they went into the dust to gather more people.

In response to this emerging need for transportation, boats of all descriptions converged on Manhattan. Some mariners acted independently, based either on what they saw or what they heard on their radios; others sought permission from the Coast Guard. Initially, the agency instructed vessels to stand by; then it asked captains to be ready for action. Eventually the Coast Guard issued a request for "all available boats" to participate in the evacuation. A mixture of activities developed, with boat operators proceeding either entirely according to their own best judgment or acting under the guidance of Coast Guard personnel or harbor pilots. Coast Guard officers and harbor pilots—the expert navigators of the waterways—teamed up to provide traffic management of boats and even of supplies, turning Lower Manhattan once again into a functioning seaport. Working together on the pilot boat *New York*, a large vessel used for dispatching pilots that is equipped with plenty of radios and radar equipment, harbor pilots and Coast Guard personnel coordinated the arrival and departure of boats around the Battery, while other personnel went ashore to help direct the evacuees. Essentially, as evacuees lined up along the waterfront, boats stood offshore, negotiated with one another for access to whatever dock space was available (sometimes tying up to trees or lampposts), and loaded passengers. Over time, some boats displayed handmade signs indicating their destinations (for instance, Hoboken or Staten Island) so that evacuees would know where to wait.

As the day went on, the boats began returning to Manhattan full of supplies and emergency response personnel. And the evacuations continued: some people had sheltered in their buildings throughout the morning and afternoon, only to attempt to leave in the early evening. One boat operator recalled five hundred people waiting along the waterfront in Midtown at 8:00 P.M. that night. Naruszewicz remembered his last run picking up evacuees:

> We went in . . . and no one came, and I had to use the spotlight to find the loading ramp. . . . I could just look up into the spotlight, catch the beam of the spotlight, and it was snowing dust. The whole horizon: concrete, insulation. It was snowing in the middle of the night. Quietly we sat there for about five minutes and then people started to trickle . . . some of them walking around like they were sleepwalking.

Others described it similarly:

> It looked like you landed on the moon. Everything was quiet. The boats are covered in this gray dust. It looked like something out of a haunted movie. That's how it looked and when you walked in the stuff, it was up past your ankles.

Two days later, there were still evacuees, people who had stayed with a friend or coworker who were now looking to head home, but most of the runs at that point were intended to bring in supplies and response workers rather than assist people looking for a way out.

Although these evacuees were often boarding vessels not designed for passengers, from waterfront locations not suited for transferring personnel, and under conditions of urgency and ambiguity, no significant accidents were related to the evacuation effort. No one knew what was going to happen next, or whether the attacks were only the first installment of a longer sequence of assaults. Under these unparalleled conditions, the boat lift required varying degrees of self-organization, improvisation, and coordination between multiple government agencies, companies, and private individuals. It was, in other words, an example of individuals and organizations learning and acting under conditions of extreme environmental stress: forming new relationships, suspending existing procedures and developing new ones, and making decisions based on ever-shifting and ambiguous information. Despite not knowing what to do, the individuals involved figured it out as they went, tackling problems as they arose and adapting their unfolding understanding to the resources available. In short, the situation was one of rapid change, with urgent needs for decision making and action in a transformed landscape.

Although the events of 9/11 were unprecedented, they fall under a more familiar category: disasters. While the specifics of disasters may vary from one event to another, people tend to respond to them in certain ways, and 9/11 was no exception. The spontaneous response of New York's maritime community to people in need demonstrates the capacities of adaptability and resourcefulness that we see often in disaster situations—capacities that emerge from particular local conditions and that are directed toward solving disaster challenges. Understanding why members of the maritime community reacted as they did requires a closer look at how communities and individuals make sense of and respond to disasters.

Human Responses to Disaster

Myths about disaster behavior abound. The widespread panic myth has been so prevalent that in years past, public officials sometimes delayed providing warning information (Fischer 2008). Kathleen Tierney (2003) describes how intelligence services in 2001 withheld credible information about a nuclear device in the city even from New York City officials. Concern that word might leak out and cause widespread panic impeded New York officials from applying their own local knowledge that might have helped them detect any unusual activity in the streets of New York and undercut their ability to act as valuable partners to federal agencies concerned about terrorism but lacking the detailed knowledge of New York's streets and law enforcement environment.

Other pernicious myths have affected decision making by public officials. Tierney, Christine Bevc, and Erica Kuligowski (2006) have analyzed how media reporting of looting, murder, and general mayhem in New Orleans during Hurricane Katrina was energized by and reinforced myths about human behavior. By spinning up indirect reports and rumors of extreme violence and lawless behavior, media reporting magnified and multiplied the few instances of actual criminal activity; in so doing, the media led public officials to believe that order had completely broken down. In response, officials devoted more resources to law enforcement than to search-and-rescue and lifesaving activities. Something similar occurred after the 2010 Haiti earthquake, in which fears of unruliness prompted a swift end to aid distribution, leaving hundreds of people in lines waiting for critical supplies that would not come. Instead, they were forced to either grab remaining food or take it from properties in the vicinity (Velotti et al. 2011). In other words, in some of these cases, measures taken to ensure security actually prompted a less secure environment (Velotti et al. 2011).

Yet another myth is that disaster survivors are too shaken to do anything and will sit around waiting to be helped (Alexander 1993). Certainly they will be upset, scared, and confused while they take it all in, but studies nevertheless show that people typically start disaster response activities quickly, making use of materials at hand for organizing search and rescue, offering first aid, and supplying food and water.

In contrast to these media-driven stories of unruly mobs, such scholars as Havidán Rodríguez, Joseph Trainor, and Enrico L. Quarantelli (2006: 91–92) have reported considerable evidence of resourceful, helping activity in the wake of disasters. For example, after Hurricane Katrina, one neighborhood group gathered people in a school and converted it into a shelter, with sleeping areas and a dining room. They also sent teams out to bring food to elderly people who had stayed in their homes. Officials might have been able to bolster these and similar activities had they been expecting them. But interpreting disasters as only occasions for mayhem, described in the language of warfare or combat, obscures aspects of human behavior that point toward different and more salutary response activities that, together, make the best of civil society. These activities include the self-help and emergent efforts generally undertaken by people in the area affected by disaster.

Rather than running away from disaster, one of the most universally observed behavioral responses in a time of crisis is *convergence*.

In a classic work emerging from Cold War concerns about behavior in crisis, scientists Charles Fritz and John Mathewson (1957: 3) define convergence as "movement or inclination and approach toward a particular point." They further describe convergence behavior as falling into certain categories, including *personal*, *materiel*, and *informational*. Some convergers are those who are returning after evacuating, some are people who are anxious about family or friends, some are there to try to help, some are curious, and a few (usually very few) are there to take advantage of people's vulnerability or overall confusion for their own profit. In protracted events, such as 9/11, mourners of the dead and well-wishers displayed another kind of convergence when they appeared at sites associated with the response (Kendra and Wachtendorf 2003). Some convergers are "official" helpers, while others are not; still others, like the boat operators, take on a sort of semiofficial status over time. Moreover, people who are officials in some circumstances may be unofficial in others: for example, the volunteer firefighters who traveled to New York from their homes in neighboring states. While the arrival of convergers can prove annoying to officials, their presence is necessary. Convergers, for instance, often provide immediate assistance to those in trouble. Others, whose role may be less obviously useful—the curious—nevertheless provide an important civic function by watching the transformation of the environment that they, as citizens, hold in common.

The phenomenon of emergence, or the development of new groups that form around accomplishing some task, contradicts the misconception that people are too stunned to react and will sit around passively while they wait for assistance. Public officials have informally settled on a period of seventy-two hours as a period of self-sufficiency, the amount of time that individuals or households should be prepared to wait before help arrives. Plentiful experience shows, however, that people will not sit alone and idle for those seventy-two hours. They will start to help one another, sometimes in small clusters and sometimes in larger and more organized groups. When people feel out of touch, isolated, or on their own, with little incoming information or contact with external help, and when they see jobs to be done, they will start to do them (Dynes 1970).

New York's mariners displayed exactly these behaviors when they responded to the emergent crisis in Manhattan. This is not to say that the mariners were unaffected by what they saw. As one person stated to us, "I'll be very frank with you. There [were] some captains who made one trip, tied their boat up from different companies and went home because they couldn't deal with it. They made one trip and they [saw] people jumping out of the windows." Similarly, another mariner described how, after being on the boat for a while, he needed to get off, if even just for a bit. He got on his bicycle, went to get a pack of cigarettes, and then came right back, but he just needed a break.

Even so, several mariners we talked with thought their ability to find some way to volunteer helped them cope with the situation. As one boat captain stated, "I think that our psychological health is probably a lot better than everybody else's, because at least we got to do something, you know? Everybody else wanted to but couldn't." Without conducting a systematic study on the direct effects of participation on coping, it is impossible to know the accuracy of such an assertion. Yet the statements from these mariners do indicate that they, at the very least, *interpreted* their participation as a positive experience. Indeed, some reported struggling with what to do after everyone was evacuated. One boat operator explained, "One of the things that was hard for me was being in the game and then all of a sudden having to be not in the game. All of a sudden you're part of it, and then all of a sudden . . . [good-bye and] take care."

Douglas Paton (2003), an experienced crisis psychologist, has written that disaster can be a source of *eustress*, or positive stress, and that suddenly getting back to normal after a period of freewheeling work, removed from regular norms and procedures, can be disturbing. One mariner responded to our question of what he did after his last run that afternoon by saying, "I don't know. I started drinking. . . . We all went over to [someone's] boat and consumed mass quantities of alcohol." We could not blame him, after what he had gone through that day.

Whatever people's emotional state, disaster research overwhelmingly suggests that people are able to act responsibly and prudently, given the information they have available to them (Kendra and Wachtendorf 2003; Quarantelli 1997; Tierney, Lindell, and Perry 2001). The 9/11 boat evacuation offers examples of all these behaviors: convergence, emergence, self-help, spontaneity, and inventiveness. Together, these responses make up the ingredients of societal continuity after crisis. Beginning with the ordinary activities of people working around the waterfront, a sequence of convergent and emergent efforts morphed into a mobile transportation system that filled in for more familiar, fixed systems that suddenly seemed to be a source of vulnerability and danger. Participants in these events, although surprised and upset, started to think at once.

What Is a Disaster?

We sat with Captain Patrick Harris, the owner of the sailing vessel *Ventura*, on the deck of his boat in North Cove, a small recreational and commercial marina on the western shore of Manhattan. In fact, it lies just to the west of the World Financial Center. Our interview took place on a day not unlike the one he was about to describe, maybe a little warmer, but with the notable exception that the Twin Towers were missing from the sky above us. He was there for the beginning of their disintegration:

> [It was a] quiet morning . . . beautiful wind, northwest wind, perfectly clear sky, quiet, and suddenly I heard this roar of an airplane. Like it was right next to us, and Jack yelled, "Look." I was facing that way and looked up and saw the first aircraft just go right in the building, and then total silence for seconds. What happened immediately after that was I kinda looked at it, and in my mind, I can actually still see the last half of the aircraft, the tail end, just tearing in the building, and then all of the windows, all the square windows like that from right to left and about four or five stories up, lit up like an arcade game, bright orange. And then I heard the sound [*he makes a puffing sound*] like you're lighting your barbeque grill and you have too much propane on before you hit the match. [*He makes a puffing sound again.*] And that was a fireball flying out of the building, big circular pattern, lot of black around it. And then it seemed to get "swoosh" sucked back in, and then it came out again, all black and smoky. My first reaction was to call

> the Coast Guard on Channel 16, which is the emergency station, and let them know. "U.S. Coast Guard, U.S. Coast Guard. *Ventura*. Emergency traffic." Something like that. They came back right away, and said, "What's the nature of the emergency?" And that's when I found myself not able to fully articulate what I saw. Because my mind still saw it, my brain registered it, but I, maybe I didn't want to believe it or something. Jack was screaming, "It was a jet; it was a jet; it was a jet," and all I told them at the time was, "There has just been a tremendous explosion at the Financial Center, trade towers, a number of stories on fire. They're gonna need backup." And that's when their protocol with the language broke down, and they said something like, "What?!?" Or, or, there was a pause, and then it was a statement back that said, "We'll alert the appropriate departments." And that was the end of the communications.

Captain Harris was very likely the first on the marine radio frequency with his startling report, describing such a shocking event that, despite its very immediacy, happening practically on top of him, he could not believe his eyes. The Coast Guard personnel on radio watch had never heard anything like that before in their lives, and after their own exclamation—we are left to imagine what they might have said off the air—they ended the conversation with an instinctively vague catch-all phrase about getting the word out. At that instant, Harris did not know—could not know—whether he was witnessing a terrible emergency in a single building or the opening of an unprecedented American disaster.

Captain Harris's experiences beg a further exploration into the concept of disaster. The word *disaster* is thrown around pretty easily, to the point that its very familiarity reduces its utility in helping us understand certain kinds of phenomena. Familiar phrases—"My office is a disaster area!"—suggest a situation of chaos, a hopeless mess. But clearly this kind of colloquial use refers to something different from a genuine disaster. Scholars have struggled for precision in defining such concepts as disaster as well as related terms, including resilience, vulnerability, crisis, and recovery. They grapple, too, with the ideas of catastrophe and megacrisis; dozens of definitions exist for each of these terms (Perry 2006). This murkiness has real-world consequences for how high-level policy makers and street-level emergency managers make decisions about how and where to direct their funding and attention. They affect how officials view the nature of a future disaster, and how they diagnose and respond to an unfolding crisis.

None of the definitions for disaster is perfect. In many ways, those that depend most on numbers are least satisfactory. Take the number of fatalities or the amount of economic loss, for instance—each has a different significance based on the scale of society that is affected. For this reason, most scholars now hold a concept of disaster that contains an implicit sliding scale, where the magnitude of the impact is considered alongside the characteristics of the society that has experienced the event, including local capacity for response. Nor can it be assumed that all hazards will have similar consequences. The decisions that people make for habitation, recreation, and commerce expose them to environmental extremes, influenced by historical conditions of social and economic development and by present efforts for planning and disaster mitigation (Mitchell 2003). For this reason, one sees different outcomes to earthquakes in Haiti and Chile, for example. The mismatch between natural, technical, and social systems (Mitchell 1990) creates conditions that make people more prone to harm. Disasters, in turn, result when the risk, which is a probability, is realized in an actual event.

Nevertheless, all the definitions involve some notion of a social system being overwhelmed and unable to meet its needs, thus requiring assistance from elsewhere. Some transformation of the natural and built environment occurs that leaves the area unable to provide its usual support for residents, workers, and visitors (Dombrowsky 1998). One particularly long-lasting concept of disaster describes it as follows:

> An event, concentrated in time and space, in which a society, or a relatively self-sufficient subdivision of a society, undergoes severe danger and incurs such losses to its members and physical appurtenances that the social structure is disrupted and the fulfillment of all or some of the essential functions of the society is prevented. (Fritz 1961: 655)

What is important in this definition is the idea of a disaster as, first, a community-wide event. Second is the disaster's effect on social systems. Plenty of specialized agencies have evolved to deal with emergencies of every sort—fire, police, ambulance services, hazardous materials cleanup companies, oil spill cleanup companies, various public health services. In everyday life, the assumption is that these kinds of emergencies are highly technical tasks and, moreover, that emergencies are official problems. Even private companies that do this work are tightly regulated and expected to work smoothly with public authorities. Disasters, though, are different. A disaster is fundamentally a *social event*. How we experience disaster is a social process.

Disasters as Social Processes

Let us imagine that, on 9/11, the towers had somehow been attacked, maybe with bombs, but no one had died. Somehow, everyone got out before the towers collapsed. Even with the same amount of physical damage, that disaster would have been much different. It still would have been huge, obviously, but the desperate urgency of those first weeks arose from shock at the scale of the event, the ties of family and friendship, and the belief that many people might have been trapped alive under the debris. Disaster, as we have seen, is a social phenomenon, rooted in our social ties and social relationships (Dombrowsky 1998). The physicality of a disaster is only part of its experience. In the moments when some physical force has removed the possibility of managing the triad of social, natural, and technical systems through normal channels, everyone suddenly takes a fresh look at the relationships between public and private institutions. The usual division of labor, especially between public and private activities, suddenly seems less useful. It blurs and vanishes as people come to the collective realization that normal systems are not, and will not be, enough. Work that used to be monopolized by a specialized agency is open to anyone, like Captain Harris, who described what he saw during that morning.

Harris got his boat underway. He heard by radio that people needed help at the South Street Seaport, on the other side of Manhattan, so he headed over there. When he arrived, though, it was deserted, so he turned around to head back to the west side of Manhattan, steering through thick smoke:

> We emerged out of the smoke—it was so heavy in there you could barely see maybe a boat length ahead. . . . By the time we emerged . . . that northwest wind was blowing the smoke, [so] I could see further down the harbor. . . . Apparently the Coast Guard sent out a message to all available boats. [He had not heard the message, though.] I saw this V-shaped formation of about a half a dozen or so tugboats charging up in this direction, and, and I remember at the time it just reminded me of the old black-and-white footage you see of Pearl Harbor. And with the smoke going over the water and with the boats in motion, and everything was really black and white. There was so much smoke you really couldn't see that blue sky. And it was actually very inspiring to see that, knowing those guys were going in there and that's where all the trouble was.

Harris and his fellow mariners were part of a community that was attempting to come to an understanding of what was happening in their own terms. In a community-wide disaster, everyone is affected, everyone is involved, and everyone takes part in defining and building up a mutual understanding of what has happened and what the community's new needs are. In an emergency—say, a water main break—a small area is affected, and there are well-established engineering procedures for handling the problem. A disaster, though, is much more open-ended—technical solutions are not obvious, and the range of community physical and social systems is disrupted.

How best to contend with those disruptions depends on how the disaster is defined by those directly or indirectly affected by the event. Towns far from Ground Zero grieved for their residents who had died, and during the protracted recovery, residents of Battery Park and the families of victims who lived elsewhere clashed over what the rebuilding should look like (PBS 2002). Was the site defined as an educational space, a site of commerce and rebirth, or a gravesite? This conflict is not unusual in disaster situations.

Disasters exist in particular social and economic contexts that affect how they are perceived (Mitchell 2006). Disaster becomes an occasion for making or creating meaning, and how we create meaning can vary, potentially opening or closing paths for intervention. Consider, for example, the aftermath of Hurricane Katrina. With reports of looting and disorder reaching her ears, Governor Kathleen Blanco of Louisiana said, of the National Guard troops who had landed in New Orleans fresh from duty in Iraq, "They have M-16s, and they're locked and loaded. . . . I have one message for these hoodlums: These troops know how to shoot and kill, and they are more than willing to do so if necessary, and I expect they will" (CNN 2005a). She was doing what many people do when confronting terrible disaster: applying combat-zone metaphors. But as Tierney, Bevc, and Kuligowski (2006: 57) have observed, "metaphors matter." That metaphor inclined Governor Blanco toward a certain set of policy choices. By contrast, General Russel Honoré, who assumed leadership of response operations in the deluged city, commanded soldiers to put their weapons down, reminding them that they were there to deliver food and other critical supplies (CNN 2005b). Instead of a combat situation, he saw a humanitarian-aid situation.

Disaster is rooted in belief, experience, emotion, and expectations about how we should relate to our environment. As real as the events of 9/11 were, their perception shaped how people acted. They had choices. They could observe what was happening from up close or far away, seeing it as a strange event from which they were prohibited, or, like Captain Harris they could enter the strange milieu of disaster.

Vulnerability and Surprise

The 9/11 attacks were a surprise not just in the usual sense but also in a more technical sense that helps us understand it in the context of other kinds of calamity. Even with danger signals across the intelligence services (National Commission on Terrorist Attacks upon the United States 2004), U.S. officials reflected the varieties of surprise that such scholars as James Mitchell (1996: 22) have identified: some imagined the possibility of attacks but regarded them as highly unlikely; some thought that attacks were likely, yet they had never happened before. When our institutions for managing danger are taken by surprise, people at the ground level cope with the consequences. We have deployed vast systems for minimizing danger (Knowles 2011): a wide-ranging enterprise of science, policies, codes, regulations, and training; a net of agencies and organizations meant to catch and hold risk. Certainly every boat operator in the harbor was taken by surprise on 9/11, even though they deal with surprises in their work every day. The way they tackled that surprise was to find the familiar and work with it: their boats, their skills, the local environment.

In its creative response to a surprising hazard, New York's maritime community demonstrated *resilience*. Resilience is a mysterious, hard-to-define quality. It is generally understood as the ability to "bounce back" (Wildavsky 1991: 77), but it also carries a strong moral flavor. Everyone, it seems, wants to be resilient; no one wants to be viewed, or to have his or her group viewed, as not being resilient. Resilience is really about the fundamental question of survival. Disaster brings us face-to-face with the elemental questions of where we are going and whether our social and technical systems will endure.

After the 9/11 attacks, and again after Hurricane Katrina, scientists and policy makers turned their attention to concepts of resilience with renewed energy. Our take on resilience is that it owes much to organizing and to building networks, building partnerships where there were none to create unexpected, but essential, connections between individuals and institutions with different skill sets. The Federal Emergency Management Agency's (FEMA's) concept of the whole community—which explicitly tries to incorporate diverse community functions—similarly tries to create resilience by encompassing the complexity of the social, cultural, and political nuances of places. Plentiful experience, however, shows us that official organizations cannot do it alone. What the boat evacuation points us toward is a strategy for thinking about the role of informal contributions alongside those of formalized disaster responders.

The palette of potential surprises is expansive, as are potential disaster needs. Our observations of some dozen of the largest recent disasters suggest that the fortitude of regular people has been a linchpin of crisis response and early recovery. We do not want to romanticize their involvement because, to be sure, excessive volunteers have challenged responders' management skills in the past. As one mariner said, "You know, there were people there more enthusiastic than useful, but you can't condemn people for being enthusiastic either." But we know that there is a place for individual volunteers, social clubs, and civic organizations that suddenly could take on emergency-related duties.

Consider the situation in California, where the U.S. Geological Survey (USGS) has updated its seismic risk estimate. According to scientists, California is overdue for a large earthquake (Chong 2009). To the extent that government agencies and citizens have been active in mitigating risk, the state will be that much more resilient. But a postearthquake disaster response will involve representatives of both government and civil society; the extent to which those domains of social life can mobilize and cooperate will affect the state's overall resilience. We are looking, in other words, at the potential of social and human capital, as well as physical and economic capital, in responding to disaster.

The maritime response to 9/11 shows a stolid and determined pragmatism that counters the geometric elegance of scripted emergency plans, with their boxes and arrows. Plans are not meaningless, but they are always beset by internal contradictions. We can never be sure how much of what is imagined for the future will be the reality of the present, which is the whole point of planning. In the absence of any plan, and in the most serious emergency to face a U.S. urban area in many years, the maritime workers of New York made a space of normalcy and turned their usual skills into a coordinated effort. Patrick Lagadec (1993), a political scientist and expert on crisis, argues that it is helpful to step back and ask what is still working in developing crises where the full extent of the trouble is unknown. The Apollo 13 flight, for example, revealed multiple system failures. Flight director Gene Krantz asked, in some frustration, "What do we got on the spacecraft that's good?" (Lewis 2015). The boat operators on 9/11 could answer that question: *they* were still good; *they* were still working.

Russell Dynes (2003), one of the founders of the disaster research field, has argued that people everywhere show considerable adaptability even under appalling, unparalleled conditions. He has based his conclusions on survivors' responses to the firebombing of Hamburg and the atomic bombing of Hiroshima (11). In Hamburg, where around forty thousand people were killed over one night of bombing, the fire brigade and first-aid workers rescued and treated more than twenty thousand people on the first night alone. Within a week, the postal service, the telegraph offices, and the banks were at least partially functioning. Similarly, in the days after the atomic bombing of Hiroshima, the military planning board and the local civic authorities did what they could to restore a sense of normalcy to the city. They placed a priority on restoring power, public transportation, telephone networks, and banks (12–13). These were among the most intense and concentrated attacks on urbanized, civilian populations in human history. Conditions must have been dreadful, but—so long as something worked—people identified ways to keep going. Some of these efforts were led by government officials and others by private citizens, but all demonstrated human capacities for coping and endurance.

Responses to disaster must involve all the institutions that normally form a community. Disasters are events of contrast and contradiction. We are prudent to establish official agencies, but we also know we need regular citizens. We plan in detail, knowing that we will have to improvise, but we also know that our improvising will be better with planning.

Sensemaking in Disaster

A disaster, as we have seen, is the outcome of many observations, interpretations, and actions on the part of people who experience a disturbance. Because a disaster is a collective experience, people make sense of their parts in it, and the whole, differently. The concept of *sensemaking* (Weick 1995: 4) is concerned with how people and organizations construct meaning in their environments. An understanding of disaster, especially as an occasion for helping and bypassing public/private boundaries, is the product of what we term *diffuse sensemaking*. In the case of 9/11, individuals and groups around New York saw something, but not necessarily the same thing, and considered what to do. Whatever pieces of the crisis they had access to shaped their awareness of the turmoil downtown and suggested different ways that they might be involved. These diffuse experiences shaped people's awareness of how they could start to untangle their initial confusion and, instead, focus on helping and response.

Sensemaking matters because individuals calibrate their responses based on what they think is happening. At some point during the waterborne evacuation of Manhattan, the Coast Guard took a lead role in managing the boat traffic. But despite an emerging narrative that says that the Coast Guard "directed" the event, it is clear that the agency did not initiate it. Moreover, even though it was active in coordinating some of the boat traffic around Lower Manhattan, that capacity faded rapidly with distance from Ground Zero. Instead, the Coast Guard *participated* with others. Mariners and waterfront workers were registering the event as an evacuation scenario that required their involvement even before the Coast Guard issued its call for "all available boats" to assist. A large number of mariners—approximately two-thirds of the people we spoke with—responded without having heard the "official" call or before the call, or they were already preparing to respond when they heard it. Audiotapes of radio transmissions clearly include vessels asking for permission to approach Manhattan, indicating that mariners, from their different locations on the water, were already poised to help and only wanted to know whether it was okay to proceed. For example, Gerard Rokosz and his staff were watching and listening from their marina in Weehawken, New Jersey, a few miles north of North Cove. About ten minutes after the second plane hit, Rokosz said, "I radioed the Coast Guard, that's 22, and I, as briefly as I could, I said, 'We know you're busy. We have a dock. If you need to get people over there or you need to use this at all, we're here.'" Operators of other vessels began to make similar preparations to assist and then actually did so upon hearing the Coast Guard's request. These instances of simultaneous interpretations of unfolding activity constitute the essence of diffuse sensemaking.

There had been no plan for a waterborne evacuation of Manhattan for the Coast Guard and pilots to coordinate, for vessels to participate in on their own, or for vessels to participate in as part of a growing structure. Instead, the effort was spontaneous and emergent, and the extent of control varied markedly across participants:

> As they [individual boats] individually responded they were asked to contact the pilot boat on channel 73, which they did. When they'd contact us we got the name of their boat, the size of their boat, the draft, how many people they thought they could fit on the boat, that kind of information. And then using that information we were able to disperse them throughout the lower part of Manhattan. (Jack Ackerman, harbor pilot, South Street Seaport Oral History)

> "We moved about 30,000 people on our six boats," says Peter Cavrell, senior vice president of sales and marketing for Circle Line. "It wasn't any kind of coordinated effort. We just started doing it." Continues Cavrell, "In its own small way, Circle Line is a symbol of New York. We just wanted to do our part." (Snyder 2001)

Alan Michael, a tour boat operator, reported having no contact with anyone else:

> Nobody was directing us that day. We all knew where we had to go, so we worked it out amongst ourselves, all the captains, who would go into which berth first or second or what have you. (South Street Seaport Oral History)

The evacuation was a pastiche of coordinated, loosely coordinated, and independent efforts that formed not so much a responsive *system* but rather a responsive *affiliation* or constellation that extended between individuals, groups, and organizations distributed over space—an affiliation that succeeded in moving heretofore unimagined quantities of people and supplies on short notice. Although some of the participants, including the Coast Guard, responded within the framework of existing contingency plans for other emergency events, such as the sinking of a Staten Island ferry, the overall operation was unplanned. Most of the participants, however, were not part of any such contingency planning; they experienced the evacuation as new and as largely or entirely undirected by any centralized control.

Studies of sensemaking in organizations are concerned with how people build up meaning—how new meaning is "inducted" (Weick, Sutcliffe, and Obstfeld 2005: 418). Was the evacuation just so obvious that it required no new thought? How did the participants move from seeing the burning towers—which could be interpreted in different ways—to converging on the waterfront? For example, some vessel operators first responded under the assumption that a small plane had struck the tower by accident, while others assumed from the beginning that a terrorist attack was underway. Many were unaware of the degree of damage, not only to the towers themselves but also to the transportation infrastructure. Few had any idea of the extent to which security measures were being implemented. And yet together, dispersed across the greater New York Harbor as they were, harbor officials, vessel operators, converging evacuees, and others created meaning in this turbulent environment as they simultaneously interacted with that environment and eventually with one another.

It did not matter to someone in New Jersey exactly what was happening in Manhattan. This is the sensemaking quality of *plausibility*, in that your understanding does not have to be exact, just good enough (Weick 1995). Boats were arriving and challenges needed to be met. The roles, the players, and their executions were improvised by the harbor community. That an evacuation was necessary, that it was urgent, that it necessitated a private and public sector response, that it might require breaking some rules, that certain supplies were necessary, that some vessels were more suited for certain emergent roles than others—these realizations constituted the elements of sensemaking diffused across participants at varying times and places surrounding New York's harbor.

Disaster experts emphasize the importance of shared knowledge and prior experience in responding to crisis (Buck, Trainor, and Aguirre 2006; Comfort 1999; Kendra and Wachtendorf 2007; Weick 1987; Weick, Sutcliffe, and Obstfeld 1999). Participants, some with shared knowledge and experience and others without, formed affiliations that turned into working relationships that met needs as they were defined. But while shared knowledge was important, so was an ability to recognize the limits of knowledge among respective participants. The participants in the various interactions succeeded in creating an emergency response affiliation that was effective in meeting overall needs. For us, diffuse sensemaking is the basis for developing an organization in the first place, an emergent multiorganizational network (Drabek 1996) that is focused on specific activities or response functions where none existed before. In going from a "flux" of disconnected, even infinite possibilities (Weick, Sutcliffe, and Obstfeld 2005: 411), by their collective and evolving definition of needs, these participants each created an evacuation scenario, one that was similar to that of others with similar knowledge even though they were in different locations. Karl Weick, Kathleen Sutcliffe, and David Obstfeld (1999: 103) have found value in organizations that allow decision-making power to "migrate" to the component best suited to solve the problem rather than operate strictly by a rigid, hierarchical, and unimaginative decision-making structure.

Boat operators, including Mark Phillips, comprehended a crisis and *took* decision-making power. Phillips ran a dinner boat line, VIP Cruises. The company had four boats carrying anywhere from 150 to 600 passengers—big boats, with staffs of servers, bartenders, and chefs as well as mariners. Their boats were chartered for private parties, birthdays, bar mitzvahs, and corporate cruises. Their clientele also included the businesses and corporations around Lower Manhattan, which chartered the boats for company picnics or holiday parties. Phillips remembered September 11 starting as a routine workday on the waterfront. With much of their work taking place in the afternoons and evenings, mornings were for maintenance and cleaning. Phillips started work at his office at the North Cove Marina, next to the World Financial Center—the same place where Captain Harris on the *Ventura* anticipated a quiet morning—and had motored in a workboat to another VIP facility in Red Hook, Brooklyn, to pick up some oil for two of his boats. Standing there at his workshop in Brooklyn, looking across the harbor, he could see the Statue of Liberty and the World Trade Center. "It was like, just a perfect view, you know?" he recalled. It was a fine day, clear with ten miles of visibility, with morning temperatures in the mid-sixties (Weather Underground 2015). While Phillips worked out back, sorting the materials for the morning's chores, one of the kitchen staff rushed in, exclaiming that the World Trade Center was on fire, that "a little plane went into the building." Phillips saw a thin plume of smoke rising and guessed that maybe a sightseer in a small plane had come too close and had crashed. "It didn't look like anything. It looked like just a little ribbon . . . a ribbon of smoke, almost like a chimney."

Many others had the same thought. On one of the many TV stations we monitored that morning, a witness interviewed by news anchors over the phone actually said he "saw" a small plane hit the building. But of course, he did not. Distance, perspective, and the complete surprise of the event overwhelmed awareness and comprehension, making witnesses "see" something that was not there. Nor did they see what *was* there: the vast, punched-out silhouette of wings spanning several floors of the tower. The phenomenon of *normalcy bias* was strong here: in an ambiguous situation, people spin facts toward the familiar and the positive, ruling out risk and danger. It had to be a small plane; what else could it be?

With no additional information and the true magnitude of the event diluted by distance, Phillips kept gathering his oil filters, oil, and tools. His next view of the ruin unfolding in Manhattan was not firsthand. One of his staff members was watching TV and called him in to look:

> Wow. From what I saw fifteen minutes prior to seeing what I saw on TV, I could not believe what I was seeing. . . . I saw what was going on, and I said, "Boy, I better get back there."

Meanwhile, Jerry Grandinetti, one of the VIP boat captains, had a different experience. His morning began leisurely, with time off in his downtown apartment following a late-night dinner cruise. He heard the impact of the first plane striking but did not know what had happened. His wife called to tell him some kind of plane had hit the World Trade Center and that he should call his mother to let her know they were all right: "I was hanging out of the window taking pictures when I heard the second plane coming in, and that was when the second plane hit. I saw people falling." He threw on some shorts and a T-shirt, little knowing that he would spend fifteen hours making history in that limited attire. He grabbed his portable VHF marine radio and rushed to the VIP office, where everything was pandemonium, everyone on the phone or yelling. He stepped outside to check on the boats—no problem there. He barely got inside the door when the first tower collapsed, darkening Lower Manhattan with smoke and dust. While at first Phillips was too far away to see what was happening, Grandinetti was too close to see. He did not know then that the first tower was collapsing; he thought the noise was an explosion: "As soon as it got light enough to see out the windows a little bit, I kind of poked my head outside in the street, but you couldn't see far. I didn't know that all this stuff on the ground was from the first [tower] coming down."

Grandinetti stepped out onto the sidewalk. He saw firefighters dragging other firefighters. He went to a supply locker to look for water to give to people who were covered in dust, gagging. He grabbed a case of water and then went back to the office; in the midst of all the commotion, he said, "There was nothing I could really do, so I went down to check on the boats again." At the docks, two firefighters asked whether he could get an inflatable Zodiac running to get down to the ferry terminals on the tip of Manhattan. They motored around the island and pulled in, where they suffered a not uncommon but poorly timed mishap: one of the lines used for tying up the boat was sucked into the propeller. The light rubber boat bobbed up and down: "I had to reach down in the water, feel it, and try to start unraveling it, but the boat was going up and then it was coming down. When it went down, my head went underneath the water. Went back up, my head came up with it." By now drenched, he dropped off the firefighters and finally headed back to North Cove but found it mostly deserted. As he tied his boat up (uneventfully this time), the second tower collapsed. Again, he did not know it had come down. It sounded to him like another explosion. He and a policeman dodged to another boat to wait out the rush of dust and debris.

From there, Grandinetti went up to the wheelhouse of the *Excalibur.* The craft was loaded with dust—dust on the deck, dust covering the windows. Captain Dennis Miano, one of the other captains in the fleet, recalled that the dust was 5 or 6 inches deep. After trying and failing to remove the dust with a squeegee, Grandinetti used a fire hose to rinse the windows. He got the engines going and, handling all the mooring lines himself, let the boat rest against the dock, holding with one line, working on it with the engine going slow to hold position: "There was nobody around, but I was getting ready to do something. I didn't know what I was going to do." A police officer on shore shouted up, "You have to get out of here!" Grandinetti called back that he wanted to but added, "But once I get in the harbor, I don't know what I'm going to do, because I don't have a crew." While Grandinetti alone could hold the boat to the dock by going slow on the engine, once he got the boat underway, he would need more help, and he would need help docking the boat later. The officer summoned some other police, who pulled up in a little boat. Grandinetti and his impromptu crew untied the *Excalibur* and steered out into the harbor, heading for the East River near the heliport, where the police knew some evacuations were happening.

Back in Brooklyn, Phillips and other VIP captains and crew were pulling themselves together. Like Grandinetti, they were getting ready to do something, as soon as they figured out what that something might be. While watching TV with his wife at home, Captain Miano even saw the *Excalibur* under Grandinetti's command motoring past Governor's Island: "We couldn't believe the cheek. How the hell did the boat get out of the damned dock?" After a series of phone calls to run through their options, Phillips, Miano, and VIP port captain Bobby Haywood—also at home in Brooklyn—eventually converged on a boatyard in Gerritsen Beach. They picked up a few more mariners along the way, including Captain Buddy DeWitt and Captain Fred Ardolino. Ardolino, who owned a competing dinner boat business but had known Haywood since they were kids, offered up his little fishing boat. They packed themselves in and started for Manhattan.

It was a rough crossing—not from the water but from the picket lines of Coast Guard and police boats that stopped and questioned them, over and over again. At least once, as one of them recalled, they were stopped at gunpoint: "The guy we had driving [our] boat was an Italian guy with a bald head. He looked like an Arab. . . . He says, 'I ain't stopping this time. We got to go.' I said, 'You better stop, or these guys will start shooting at you.'" Even at this early stage, they expected some sort of a backlash and official reaction made through the lens of ethnicity (Peek 2011). At each stop, they explained who they were and what they were trying to do and either persuaded the officers to let them pass or at least convinced them to call a superior for permission: "It took us an hour just to get into the harbor. We just kept telling them, 'Get us some clearance to let us through.'" Finally, one of boats they encountered—it could have been a Coast Guard or a police department vessel—called another boat to escort them into North Cove.

Once there, they found the place heaped with the detritus of catastrophe, smoothed over with drifts of toxic, glittering particles. "It looked like shiny moon dust," Miano said. They wanted to get underway, as Grandinetti had done previously, but no one was around. Phillips and Haywood could handle the *Romantica*, but Miano's *Royal Princess* was a much bigger boat. Miano still had one of the boat operators from Brooklyn, but for a boat that size, he needed more crew. He ran into Harris, of the *Ventura*, who jumped on board and lent a hand for the rest of the day. With the help of an unlikely pair—a couple of foreign tourists standing around, neither of whom had any maritime experience—Miano eventually pulled the *Royal Princess* out of North Cove.

For Phillips, Grandinetti, Miano, and Haywood, the next hours passed with multiple boatloads of evacuees heading in all directions. The first load of people went to Brooklyn, with Grandinetti. It was just automatic, Phillips said, making runs back and forth. The first time they were instructed by the Coast Guard, but after that they ran at their own discretion. Sometimes they went to Liberty Landing, near New Jersey's waterfront science museum; sometimes they went to a marina a bit farther north. Once they were asked to pick up a contingent of police officers from the East River side of Manhattan.

Phillips and his colleagues had, essentially, taken on new jobs as emergency responders. Their roles shifted and evolved with changing needs ashore and at Ground Zero. They did not just carry passengers: "We ended up bringing body bags. We were bringing medical teams, dogs," said Phillips. They transported hundreds of cases of bottled water. Because the dinner cruise boats had lots of lights, the

New York Harbor pilots who were helping coordinate boat operations asked Phillips to keep one of the boats going at night. That evening, around six or seven o'clock, the boats were loaded with gurneys to make a triage and treatment area. And since they already had a lot of food on hand for the day's canceled excursions, VIP's boats also served as a dining hall and rest station for weary firefighters and rescuers. They solved problems as they arose, fitting the supplies and skills they had to needs that developed.

All of these activities stemmed from moment-to-moment interpretations of what was happening in the environment and what could be done with the people and resources available. Each step yielded a new inspiration. If hazard and disaster stem from a mismatched interaction of natural, social, and technical environments, then responses to them must involve investigating and reworking those features to rectify the mismatch. While these responses may not necessarily be conscious acts, people in a disaster zone must learn to rearrange those pieces. One mariner said, "That was the thing that happened: people would come in and they'd lend their expertise, whatever [it] happened to be. . . . Most of the time, they were schlepping stuff around, but then all of a sudden, somebody would come up with an idea, and if it sounded like it was decent, we'd go with it."

Resiliency depends on knowledge of the environment. A community can either understand enough about the potential hazards it faces to design an impervious system of protection, or it can be so well informed about its environment that it can regain its equilibrium when disturbed. But since it is difficult to anticipate everything, communities need to be able to improvise as well as plan ahead (Weick 1993; Wildavsky 1991). Theatrical improvisers exercise skills that allow them to perform skits and routines spontaneously (Wachtendorf 2004). They are making things up as they go, but they know which principles to pull together. They know how to make use of props and cues and the environment closest to them, the theatrical space. Instead of following a scripted plan, improv performers match what they know and what they have at hand to generate a memorable performance. They create the "platform"—in other words, an understanding among themselves of who their characters are, what they are doing, and where they are doing it—and then signal to the audience in a way that allows the audience to "see" and understand, even in the absence of set and props. In the same way, emergency responders—official and unofficial alike—generated meaning, understanding, and ad hoc organizations from certain cues in their environment. They built, populated, and represented to one another many platforms all around the harbor.

Disasters do not have boundaries. Emergency managers' habit of referring to the "big picture" is quite dangerous. As real as they are on the ground, disasters have a contingent, ephemeral quality. Any one person's "picture" is by necessity limited, provisional, and transient. Captain Harris had one picture, as debris fell down around him; Grandinetti, covered in dust, had another. All would soon be involved in the same events, but at a remove of distance and awareness that produced different realities for each of them. Yet somehow, they each came to a similar, albeit diffused, sense of how to respond.

The distance from Ground Zero to North Cove, where much of the waterborne action took place, is about 350 yards, or about as far as the towers were tall. Yet from the standpoint of managing an unfolding calamity, those places were in different worlds. If no one in that amount of space had the big picture, how was it possible for an expanding system of people, places, and equipment across the harbor to grasp it? Or, for that matter, across the country? The picture, instead, was formed, envisioned, and created by everyone together, either at the same time or in sequence. By the time official responders could begin to react, other individuals had already become involved. As with starlight, emergency management officials' "big picture" was an image that was produced quite some time ago. By the time they saw it, the real situation had moved on.

This is the idea of disaster as socially constructed. A disaster is not a thing or a place: it is an idea, based on effects and interpretations. What, for example, was the disaster area for 9/11? The sixteen acres of the World Trade Center? Lower Manhattan? The area south of Canal Street? All of New York City? Any of these were reasonable answers. This philosophical question has practical significance. It means that the shape of a disaster changes constantly, resisting scripted plans and procedures. In time, of course, the metamorphosis slows, and additions to the disaster system become less abrupt. Officials become familiar with the changed environment, and plans can once again align with reality. In disaster events that are familiar or expected, the period of creating and improvising can be brief. During novel or surprising events, the improvisational time can be lengthy, ranging over months or even longer.

Almost always, people can help one another and take whatever steps they can to stabilize damage, rescue survivors, and begin to care for themselves and their neighbors, even before "official" help arrives. This characteristic of human behavior, which has been consistently documented in disaster research spanning over a half century, leads us to rebut a larger myth about disaster: that people must transcend their identities or somehow become larger than themselves, larger than life, to react during a disaster. We see this myth in the easy use of the word *hero*, which by its very meaning sets up an apparent divide between those who do and those who do not—the active givers and the passive receivers. No doubt people take some unusual actions in disasters, but the story of the waterborne evacuation of Manhattan shows how many of those responses were the projection of people's ordinary capacities into the uncertain disaster setting. We shall discuss how the everyday work of people, using their everyday skills, is the backbone of disaster response—the heavy-lifting muscle of getting a lot of work done, of helping our neighbors and rebuilding our communities. Our identities in "regular life" extend into our responses to disaster.

2 /

We Did What We Had to Do

Identity, Ethos, and Community in Action

Scott Shields used to run a company that performed marine rescues in New York Harbor, in cooperation with the Urban Park Service. He had written a handbook on safety for swim races, been involved with the U.S. Coast Guard Auxiliary, and taken plenty of logistics and emergency management courses over the years. Shields was in Connecticut at his sister's place when the first plane struck; he arrived in New York City less than an hour after the second plane hit. "I always keep a bullhorn in my vehicle," he said. "Turns out that was one of the most important tools I had that day." As the disaster unfolded and the day went on, he soon noticed a number of responders having a difficult time negotiating the gridlock on the West Side Highway. Under the circumstances, walking to and from the area carrying heavy gear and equipment just seemed too much to ask. Shields eventually suggested to personnel on the Coast Guard's buoy tender *Katherine Walker* that boats could be used to ferry response workers back and forth from uptown above dense traffic gridlock down to North Cove, right near Ground Zero, saving time and—more importantly—the energy of the exhausted workers. Shields said they did not even ask his name but told him it was a good plan. He then encountered another high-ranking official who resisted the idea, telling Shields, "We've been told that the terrorists are coming in by water and we can't rely on [boats]." Bullhorn in hand, Shields pressed on. Eventually, enough people sided with his perspective that the firefighter-ferrying operation got underway.

Another resourceful participant was Adam Brown, a diver and underwater welder. He did underwater structural inspections and had formed the not-for-profit Working Waterfront Association. Brown was at a port safety conference in Washington, D.C., that morning and was able to catch a train back into New York City on September 12. As soon as he got home, he grabbed his welding equipment and headed down to Ground Zero, intending to help burn and cut metal. It was difficult to get in, so he instead headed to Pier 63, on the west side of Manhattan, where he caught a ride on a boat bringing supplies. His initial response to the rescue operations on the pile of debris from the towers' collapse was that it was disorganized, but he worked for a while on a bucket brigade until he realized the magnitude of the task confronting them: "I was with the bucket brigade probably for a couple of hours when I'm thinking, 'We're not doing anything here. This is, you know, this isn't happening.' And I bumped into a friend of mine"—Shields and his search-and-rescue dog Bear. He knew Shields from the swim races in the harbor, where both of them had coordinated search-and-rescue planning. He worked with Shields and Bear, combing the debris until Bear needed a break. Brown initially thought of himself as a welder and a medic with search-and-rescue experience, and this background influenced the activities he became involved with, at least at first. He then found a bucket brigade. Then he ran into an acquaintance—Shields—doing search and rescue in a different way. It was a dynamic process of asking what needed to be done, asking himself where he might be most useful, and then repeating that again and again. He even hooked up with a harbor pilot for a time and helped with vessel traffic.

Within twenty-four hours after the attacks, aid stations had been set up, volunteers had arrived to help feed people working in the area, and Boy Scouts were helping unload items. An area just south of North Cove Marina took on the appearance of a military supply depot as it was converted into a space for inventorying and storing supplies. The American Red Cross and the Salvation Army were involved, as were many other organizations and unaffiliated volunteers. As one person put it, "It was the most unorganized yet organized thing I've ever seen go on. [Someone] would show up, get on a line, start carrying boxes, [and then] leave and do something else. People just did what they had to do."

Time and time again, we heard that same statement, or variations of it: people just did what they had to do. As one person put it, "I don't think there was much commanding to do. There was a job to be done, and people did it." Supplies, too, seemed to magically appear: one person we talked with recalled a time when he put out a call on VHF Channel 13 for eight-inch crescent wrenches to open acetylene bottles. Soon, "every tugboat [that] came by tossed us an eight-inch crescent wrench." The crescent wrench example seems to have been especially memorable: three mariners recounted this incident to us from different perspectives.

Such sentiments are not restricted to the events of 9/11, as we have heard variations on this phrase regarding historical and more recent times of crisis. Take, for example, the immediate aftermath following the August 2011 shooting rampage on a youth summer camp on the Norwegian island of Utoya, when private boat operators spontaneously converged and worked alongside formal responders, even while the killer—a political extremist—continued to pose a threat. A *Telegraph* interview with one of the boat operators—German tourist Marcel Gleffe—cited him as stating a phrase reminiscent of what we heard from 9/11 boat operators: "I just did it on instinct. . . . You don't get scared in a situation like that, you just do what it takes" ("Hero of Utoya Island" 2011). An anecdotal account by Steven Weintraub, a consultant to the Port Authority on the preservation of 9/11 artifacts, illustrates a similar sentiment. A number of years ago, Weintraub had a conversation with the son of the owner of a fishing boat used to evacuate Danish Jews to Sweden during the German occupation. At the time, the vessel was being donated to the collection of the U.S. Holocaust Memorial Museum. Like the New York Harbor mariners and like the German tourist who responded to the Utoya shooting, the Danish man stated emphatically that they did nothing heroic with respect to the evacuation; rather, they just did what they had to do (Steven Weintraub, pers. comm.).

We can take it as a given that people want to help. But desire alone is not enough to make them effective or capable of coordinating complex activities. How to bring together diverse people, who may or may not be familiar with one another and who have little information, is the enduring puzzle in disaster science. Disaster responses are characterized by separate activities, guided by certain overall goals and sensemaking diffused across space. Although coordination is usually understood as working together, it can also be understood as working separately, as in not getting in someone else's way. Coordination includes a capacity to form boundaries to avoid harmful interference with other actors, especially when information is lacking. So how did boat operators, waterfront workers, harbor pilots, Coast Guard officers, and others figure out what to do? How did they figure out how to work, together and separately, on such a large scale?

A key element in understanding how people made sense of this event and their abilities to respond to it is *identity*. Many of the participants referred, directly or indirectly, to their identities. Identity includes more than merely who people are—it includes their skills, experiences, and occupational backgrounds. We know from previous studies that identity is a central determinant of how people respond to crisis. For example, Karl Weick (1993), an organizational psychologist, emphasizes identity in his approach to resilience in his landmark reinterpretation of the Mann Gulch wildfire incident. Weick studied the events that led to the deaths of thirteen smoke jumpers in a fast-moving grass fire in 1949. Weick found that what people see and interpret owes much to who they are. At a key moment during the Mann Gulch fire, the boss shouted for the crew to drop their tools and take shelter in the ashes of an escape fire that he had lit. But dropping their tools would have meant dropping their identities as firefighters. "Who are we? Firefighters? With no tools?" (635). This particular group of smoke jumpers was new, with no experience with their boss or with his then unorthodox backfire; his command seemed strange and disorienting. Weighed down by their tools, they were overtaken by flames, and they perished.

As we have seen, sensemaking is concerned with how people and organizations construct meaning in their environment. Weick (1995) has identified many components to the sensemaking process, but one of the most important is that it is "grounded in identity construction" (18). Actors first make sense of their identities, which not only provide patterns for interpreting events but also can be shaped by events. As Weick puts it:

> Once I know who I am then I know what is out there. But the direction of causality flows just as often from the situation to a definition of self as it does the other way. (1995: 20)

Tools and skills are vital signifiers of individual identity that enable group cohesion (Weick 1993). Skill, knowledge, and philosophical outlook are used to distinguish who is *in* or *out* of a particular group, a phenomenon seen in many professions, such as among the diverse members of the medical field, who often argue about professional standing (Miller 1998). Skill, too, affects the kinds of cues that people extract from their environments and helps form the evolving awareness of events that collectively constitute diffuse sensemaking.

Decisions about how to improvise were not always clear during the early hours and days of the 9/11 response. Most of the time, they reflected the perspective of the organization, the individual, the person's mission or that person's interpretation of what he or she was seeing around them. In the crisis of the moment, some people were open to new ideas; others were not. Whose perspective was eventually implemented depended on rank, the number of people in the vicinity who shared the same perspective, or the willingness of the people involved to fight for their vision, a street-level politics of negotiation, persuading, and bluffing. Almost always, it depended on identity.

Emergent Leaders

The name *Ken Peterson* popped up again and again as we spoke with mariners. We were familiar with his name and reputation. He is featured in the South Street Seaport Museum's Oral History. His image and words are on display at the museum's exhibit, and abstracts from his interview are included in the accompanying volume, *All Available Boats*. He looks the part of the veteran tugboatman. He himself admitted that some may have grumbled about his style and the attention directed his way, and, indeed, he seems to have been the most conspicuous of the participants. Still, no one we spoke with denied his extensive knowledge, his technical skill, or his sound judgment.

Peterson was a graduate of Maine Maritime Academy and had cleaned oil spills for about twenty years in New York Harbor. He had worked with almost every company in the area. As port captain for Reinauer Transportation, an influential tugboat company in the Northeast, Peterson was the one ashore in support of seagoing operations, whose job was to be sure boats had the necessary crew, equipment, and fuel to operate efficiently and in compliance with regulations. He developed his knowledge for this over years aboard tugs. By the time of 9/11, Peterson had a broad palette of skills as a boat operator and a fleet coordinator—just the knowledge needed to act as spontaneous port captain for a spontaneous fleet in an improvised seaport.

Peterson described to us what he saw from his location on the other side of the harbor:

> When the first plane hit, I was doing an inspection on the *Morgan Reinauer*, one of the tugboats for the company. I was standing there looking out the window while talking to the captain, and I saw this plane . . . and was like . . . what the heck, is this a movie or something? I'm, like, something's wrong. And I looked and I looked again, and it was still there. And then it came over the radio, something happened. So I went up to the main office, and I grabbed Bert Reinauer and the head dispatcher at the time, and we came out on the front deck of the property. And we were looking out over the water and saw the second one hit. And they took all the captains, and, uh, we went inside. There were four tugboats at the dock there.

According to Peterson, one of the shoreside officials thought they should stay put:

> His first comment was "We don't want to go [anywhere]. We don't want to do [anything]." I was like, "We gotta go. We're

> here. We do oil spills. We do cleanups. We help everybody else. Why don't we go and do this, too?"

They brought in other staff, including company officers and owners, talking it out:

> So, we go up to the captains and, uh, said, "Hey, we don't know what we're gonna see, we don't know what we're gonna do, but we're gonna go. And I wanna be the lead, and we're gonna go for us." They're like, "We'll talk to our people," and I said, "Volunteers only. If you wanna go, you're welcome to go, but the owner's lettin' the boats go."

At the time, Peterson said they had not heard the call for all available boats, but one of the tug captains reported the call for help at some point during the deployment.

Kevin Tone also worked for Reinauer, as the safety director. Tone had a background with the Coast Guard, and he had worked with Reinauer for the past eight years. He had a similar recollection, although it differs in timing from Peterson's:

> I would say, and this is the unusual part, it took maybe ten to maybe fifteen minutes after the second tower had [fallen] that we grouped together at the office . . . and said, there must be something we can do. I think somebody advised at that time that the . . . Coast Guard, VTS, or command center had come . . . [over] the radio . . . [requesting] immediate assistance or support from any vessels in the area. . . . So we started sizing up what we can actually get underway. You know, at the time . . . the company is a commercial company. You have contracts committed. . . . The tugs totally take barges full of oil and gas to various sites for our customers. But we knew what was happening, that probably all activity in the port would cease until they knew what was going on. We weren't sure that the attack had ended, who knew. . . . what else was [coming], but they asked for help. . . . So we lined up several vessels that would make the transit over there. We didn't know what we were

gonna do, but we knew that, you know, obviously the impacted site is Manhattan. You know, the idea was to sail over there. So Ken Peterson, being port captain, was going to head up the little flotilla. They would get underway, the armada.

Several points jump out from Tone's statement. First, the staff and owners of Reinauer had options to consider: how many boats to send; how to define what counted as an "available" boat, given their other commitments; and what they might face in terms of overall traffic management in the harbor. Moreover, they did not know exactly how they were going to proceed. They were not responding to a scripted mission: they would be seeing and enacting, adding their ongoing sensemaking to that of others.

Tone's role as safety manager made him think to have the tugs bring along extra flotation devices, water, oxygen, and first-aid kits. His recollection, in contrast to Peterson's, was that they heard the call for all available boats while still on the pier at Staten Island, before they removed the towing gear for safety purposes. This discrepancy is not unusual. Some details are simply difficult to recall; moreover, it is entirely possible that some people within the same company or even on the same vessel might have responded to the call, while others were unaware of the situation until later. Either way, they were moving in the same direction, toward the same path of action.

They set up to get underway in four boats. Peterson was on the *Franklin Reinauer*, accompanied by the *Janice Ann Reinauer*, the *Morgan Reinauer*, and the *John Reinauer*. They pillaged their storerooms for whatever equipment came to hand that they thought might be useful. As Peterson indicated, "We thought that we needed to have blankets to wash the people, water, clothes . . . respirators." Among the crew, there was an eye for what might be needed in an oil spill, and they extended this logic to what might be needed in this emerging environment. They also brought first-aid and medical supplies, under Tone's direction and foresight as a certified EMT. The group cleared away as much of the towing equipment, wire, and other hardware as they could to eliminate tripping hazards and other potentially dangerous obstructions.

Explaining why, Tone emphasized:

> We weren't sure what we were looking at. We knew that we probably would be evacuating people, and we assumed there would be people [who] wouldn't be taken to hospitals or treated at medivac sites . . . [u]pset and that kind of stuff. But you never know. . . . You can have somebody go into shock or something on their way, you know. Some of these people may never have been on a vessel in their life, and suddenly they're on a commercial towing vessel. It can be a little intimidating. . . . It's not your Circle Line cruise boat. Not that they are unsafe, but they are a very different layout, obviously. There are [no] nice bench seats to sit on. It's not the Staten Island Ferry.

The ad hoc Reinauer flotilla approached the island. The captains and crew could see the smoke and dust. For Peterson, "It was, like, 'What are we going to do?'" Peterson and his four boats stood offshore, lined up with other tugs and smaller vessels, but he said that no one was approaching the island. People were standing around on shore, he said, yelling and waving their arms while smoke poured from Manhattan. He radioed the Coast Guard, "We'd like to have permission to get on the island to move these people." With permission, the tugs nosed up to the waterfront, pushing their bows against the seawall. "Everybody was so close together. . . . The boats were just literally stacked one on top of another," Peterson said.

The scene was not exactly out of control, but it was haphazard and kinetic, with people climbing over the fence that ran along the seawall, climbing onto boats. As Peterson recalled:

> And I got off and they made a comment that I looked like a commando, [be]cause I took the boat's radio, and I took another radio, and I had my cell phone. So I had the two radios like this. I'm standing there with all these people, and they're saying, "We gotta go! We gotta go!" I was like, "Everybody just get on the boats," and then to the boat captains, I was getting on the radio . . . was telling everybody . . . no more—maximum—than one hundred people on a boat.

Photographs of the shoreline show people walking everywhere, and sailors and waterfront workers described people climbing over fences and crawling over handrails and over obstructions. Over the radio, Peterson started to dispatch boats to different places. With the boats assigned, Peterson and others turned their attention to the queues of people pressing for transport: "We got on the wall, and we told everybody to line up, and we said, you know, 'People going to Staten Island, you go here. You're going to Jersey City, you go here.'"

In the waterborne evacuation, participants experienced several identities that shaped their responses: for example, some as New Yorkers or northern New Jerseyans, as Americans, and as skilled boat operators with capabilities that could be useful. Reinauer's tugboat operators did not usually think of themselves as running a ferry service but took on the role when it was presented to them, because they also saw themselves as people who tackle trouble. Hence their understandings of unfolding events and how they might contribute were based on their multiple identities, including identities that might actually be created through the event itself. For example, Glen Miller owns a launch company that serves the ships in the harbor, doing such jobs as transporting people and equipment to and from ships or cleaning up oil. His identity was central to how he made sense of what needed to be done:

> My knee-jerk reaction is to run to fires instead of running away from fires. . . . It's, you know, just the type of business I'm in, being in [the] oil . . . cleanup business. . . . My knee-jerk reaction was to start getting people . . . and equipment, the vessels, together, and—and [I] sought to assemble a group of people [who] would be willing to go.

In the maritime community, the imperative toward rescue is strong. In fact, shipmasters are compelled by statute to come to the aid of vessels in distress at sea if they can do so without serious danger to their own vessels. This duty means a certain amount of risk taking is *required* in the maritime community, a topic we return to in Chapter 4. One person we interviewed said, "You render assistance to any other boater in distress. That's one of the laws of sea." Another said, "If you hear a distress call and you're nearby, you go." We asked whether the mariners had thought about how the tradition applied in this circumstance, since this was a city in distress and not another boat. Repeatedly, we were told some variation of the following: "The tradition's already there, you know, the fabric is already made. You know, it just needed to be applied to the situation."

The pull of maritime tradition, with its responsibility to rescue, seems to have been strong enough to act as a sort of *master status* (Hughes 1945: 357). A *master status* is a dominant identifier that works to determine a person's position in a given society. Pamela Hepburn's experience, and how her self-defined master status of mariner drove her to the waterfront, illustrates this idea. A tugboat captain by profession, Hepburn piloted her twenty-six-foot motorized whaleboat during the evacuation and served as crew on the *John J. Harvey* starting on September 12 (about which more below). In an interview for the South Street Seaport Museum, she described the hectic morning. It was Election Day, and after dropping her daughter off at school on Twenty-First Street, she was on her way to help out with the polls in her neighborhood when the first plane struck. Not quite knowing what to do, she went back and forth between the polling station on Greenwich Street at a nearby school and her home. Eventually, she helped everyone at the school, where voting was supposed to take place, evacuate. She made her way up to Twenty-First Street to pick up her daughter, and they found themselves with their friends the Kreveys at Pier 63.

The entire day, Hepburn had made gestures at helping, be it at the polling station where residents were no longer concerned with voting or in working with others to evacuate the school children. But she also knew a boat evacuation was getting underway. Hepburn left her daughter with her friends at the pier and bicycled to her boat on Pier 25. Soon, along with a few others, she was helping by transporting fourteen people at a time slowly up to Hoboken.

Why did Hepburn gravitate to the waterfront and, ultimately, to her whaleboat?

> [I needed] to do something . . . [and] the only thing I can do is the boat thing. I'm not very good at anything else so. . . . It

> seemed John [Krevey] . . . orchestrat[ed] an evacuation from his pier [at] Pier 63 so it seemed the obvious thing to do was to put whatever boat I had . . . into service. You know, our contribution was tiny but it gave me an opportunity to do something. (South Street Seaport Oral History)

Clearly she could do many things other than "the boat thing." But she was a mariner, and that identity called to her when she was figuring out how to help.

Cynics often worry that emergency officials will not show up to help but will instead abandon their roles in the face of a catastrophic event. But here was a disaster that involved two direct attacks on the city, with rumors of more attacks on the way. Not only official first responders but also private citizens converged on the impact and response sites. Research on disasters has shown that, provided they know that their families are safe, people do not typically abandon their official roles in a disaster (Trainor and Barsky 2011). Hepburn's experience not only reinforces that finding but also shows its applicability in circumstances where a response role is not necessarily expected. She did not have to help, but she still did. Hepburn knew her daughter was safe, initially because she was at a school a good distance away from the towers, and later because the daughter was with friends. Similarly, Lincoln Harbor dock master Janer Vasquez drove his wife and children home before heading out to help with the evacuation, and he checked in at home throughout the day. We heard of a few instances where crew members opted to not participate in the evacuation, making the understandable decision instead to check on the whereabouts of family members who were in or near the World Trade Center complex. Again, these were not official responders with a formalized obligation to respond. They were private citizens, acting in a way consistent with the research by needing to check on the safety of their families before becoming involved.

The boat operators strongly felt their identities as mariners, so much so that at least some of them sidelined other possibilities for helping in different ways. Identity was part of the spark that set them thinking about what to do. In the waterborne evacuation, the imperative toward rescue created an existential *ignition* of diffuse sensemaking. In contrast to Weick's firefighters, who refused to drop their tools when doing so would have threatened their identity, the mariners registered their identity and *then* picked *up* their tools. (See Beunza and Stark 2003 for a slightly different reversal of Weick's imagery.) Identity can pull a group together, then, and not just explain its demise. And that identity also brought a knowledge of the technical environment—the skills and habits of operating vessels under demanding conditions.

Tools, Experience, and Flexibility

The *John J. Harvey* was a fireboat in service for the Fire Department of New York (FDNY) from 1931 to 1994 (Fireboat.org, 2015.). The first large modern fireboat built in the United States, it had the capacity to pump eighteen thousand gallons of water a minute. As a historic cornerstone of the maritime contribution to firefighting in the city over the twentieth century, it drew the attention of a group of maritime enthusiasts who purchased the fireboat at an auction in 1999. By that time, the boat was in a state of great disrepair. The group set about fixing it up and eventually turned their informal association into a not-for-profit organization.

Huntley Gill, an architect and historic preservationist, was in the cadre of mariners who had been restoring the fireboat. He was one of the lucky people with a job that allowed him to keep his own hours, and he had been up quite late the night of September 10. The fireboat had been involved in a water display for a Marc Jacobs fashion shoot, and when Gill finally arrived home, he had unplugged his phone before going to sleep. When he woke up the next morning and turned on the television to check the weather, he thought he was watching a movie. It took him a moment to realize, "My God, it's the news!" (South Street Seaport Oral History). He checked his phone messages and found about twelve messages from other people involved with the *Harvey*. He soon connected with Chase Wells. Wells, one of the owners of the *Harvey*, had been able to get through to the dispatchers in the fire department: "We've got this boat. Is it useful?" The dispatcher told him to just head down to the boat, and he met Gill at the *Harvey*. Wells had also called up Tim Ivory, his chief engineer, at home in New Jersey and told him to get into the city—which was easier said than done. After grabbing his *Harvey* T-shirt out of the laundry basket, Ivory headed to the George Washington Bridge. By then, like all bridges and tunnels heading into Manhattan, it was shut down. A police officer asked whether he had any ID. Ivory pointed to his T-shirt and said he was on his way to help.

Ivory was directed to a Port Authority building. Some people discussed putting him on a ferry, but that would take some time. Meanwhile, the person he was talking with got a call. The first tower had come down. Ivory got back on his motorcycle and headed to Grand Cove Marina a couple of miles south of the bridge, still on the Jersey side. There, he ran into Pete Monte, an acquaintance and fellow mechanic (Ivory was a former firefighter but now a freelance mechanic). Monte called out to a guy whose boat he routinely fixed, "Hey Frank, can you give this guy a ride down to the city?" Ivory was on his way (South Street Seaport Museum Oral History).

By now, the FDNY knew that it needed all the help it could get, even if it was not entirely clear what could be done with an antique fireboat. Wells somehow found his way into the city, joining a police motorcade along the way. Others came in from around New York. Andrew Furber, a welder and deckhand, and Tomas Cavallaro, a volunteer and artist living near Pier 63, where the boat was docked, joined Wells and Gill. Ivory finally arrived, too. Without much discussion, they were headed south. They knew that the towers had collapsed but had not heard any additional information over the radio. They essentially headed toward the World Trade Center area and waited. As one of the fireboat operators said, "We figured we'd be out there, [and] we'll see if we could be useful." Soon they started evacuating people from the waterfront. The *Harvey* took on about twice the number of passengers it usually boarded and transported them along the Hudson River, away from Ground Zero, to Pier 40.

Had it not been for a chance encounter with Thomas White, a lieutenant in the FDNY marine division, the *Harvey* might have run passengers all day. White was at home when he heard the news and immediately started making calls to figure out how to get to the site. It did not take long for him to realize that all the roads into the city were closed. Fortunately, White had his gear with him. He and a neighbor hopped into his truck and headed out to Tower Ridge Yacht Club, along the Hudson, where he kept his boat. Two others joined them. They were all north of the George Washington Bridge when the second tower came down. The radio communication was too chaotic, so they simply shut it off. When they arrived at North Cove, White's neighbor, a photographer, headed out on his own, while White and the other passenger from the yacht club took the boat back up the river. At some point, White took a walk up to Liberty Street and saw the *John J. Harvey*.

A short time later, when firefighters were telling White that the water main was crushed and they were not able to get lines in to supply water, he remembered the retired fireboat. White put in a call to fireboat pilot Jim Campanelli on the *John McKean*, from Marine Company 1, asking Campanelli to find out whether the *Harvey*'s pumps worked. About that time, the *Harvey* was passing the *McKean*. The *McKean* flagged down the boat and asked the crew to come help with the fire-suppression operation. When the *Harvey* arrived adjacent to the World Financial Center, White was there to greet it. Gill recalled his saying, "'I assume you're here for the duration.' And we [the crew of the *Harvey*] kind of looked at one another, and went, 'Well, yeah. Sure.'" Their involvement, in other words, was quite sudden for Gill. As he noted, "Literally, you know, I was in bed an hour and a half before" (South Street Seaport Museum Oral History).

At first, the FDNY proposed putting its own crew on board, but the crew of the *Harvey* resisted. That was fine with the FDNY. White gave the *Harvey* a fire department radio and unofficially designated the boat Marine 2: "When you hear Marine 2, you answer." It turned out, unbeknownst to White, that Marine 2 was the *Harvey*'s designation before it was decommissioned. Later, when White explained the situation to a chief, his superior said, "Good heads-up thinking." The fact that he did not actually have the authority to designate a private boat "Marine 2" was not, for the moment, important. Disasters are often open spaces in which people can find opportunities to take on emergent leading roles based on what they can do rather than on their routine titles or ranks.

Other boat operators described similar interactions with the FDNY. Despite the huge losses to their ranks, "everybody kind of went into autopilot. I mean it was—it was really cool to see. People just did what they had to do and . . . there was an enormous sense of efficiency to it." White recalled the emergent structure at the site: "Yeah, whoever was there . . . just whoever stepped up to be the boss was the boss. That's pretty much what it was, you know? Like, I had a chief helping me. He was [a few] ranks above me, and he was pulling hose. And I was saying, 'OK, go over there. Put this through there.'"

Ivory had walked away from the fire department years before. "It's funny how your life kind of [comes] full circle. . . . [A]ll this crap that you learned when you were younger . . . you manage to put it to use in this one event. . . . If I hadn't been a fireman, I wouldn't have known anything about hose threads and, you know, hydraulic pressures and all that." But being a mechanic was also important to his ability to piece equipment together.

As Gill remarked, "We're not first responders, we're—we're a bunch of boat enthusiasts here . . . [who] happened to have a national historic registered boat that proved useful." And yet, even so, the *Harvey*'s crew absorbed some firefighter identity, even as their boat kept vestiges of its former heritage. The crew had not directly imagined themselves as central to a catastrophic fire-suppression operation. At the same time, it was not as though helping out in some way had not crossed their minds. They had been in contact with the FDNY over the years, just letting them know that the boat was there if it were ever needed. The crew was also sure to have supplies on board. As Gill recalled, "Bob [Lenny] had always said, 'Gotta have a little bit of hose. . . . Go by a boat that's burning, it'd be really embarrassing not to have a little bit of hose.'" It was an offhanded remark, but it speaks to the pride and preparedness of these mariners. They were on a fireboat, after all, and were invested in ensuring its reputation built on years of fire service.

Retired firefighter Bob Lenny, who was among the *Harvey* enthusiasts, used to pilot the fireboat before it was decommissioned. Other part owners of the vessel suggested that he had "[sniffed] them out" when they purchased the *Harvey* out of concern about what would happen to what he thought of as "his own" fireboat. Since that time, he had become one of their most active volunteers. Someone recalled Lenny's saying at some point, "Don't sell the *Harvey*, because one day you're going to get caught with your pants down, because New York City needs to have the battleships," as he referred to them. Experienced "helpers" frequently imagine ways in which they and their tools might be useful in the event of an emergency and often have knowledge of tools unknown to emergency-planning officials. Here, the *Harvey*'s crew were dedicated to preserving old-style firefighting capacities. As Lenny explained, the city wanted smaller boats because they required less manpower, but "when they need[ed] water . . . there's nothing else that can do it like the . . . [older boats]" (South Street Seaport Museum Oral History). The *Harvey* crew knew that those capacities would one day be needed: that there would come a day when old-fashioned capabilities to simply pump water would make a difference.

Multiple Paths to Helping

John Krevey had been an electrical contractor for twenty years before operating Pier 63. After the first plane struck, he and a few other friends headed down to North Cove in a small boat to take a closer look. He saw the *Harvey*, of which he was also part owner, already preparing to evacuate people. He said, "That was the first thought put into my head by anyone that there was going to now be an evacuation effort; it hadn't even crossed my mind." Returning to Pier 63, he began organizing a waterborne evacuation effort from there. Soon small boats were showing up to transport ten, twenty, or thirty people at a time, either across to New Jersey or to another pier along the Hudson, where people could catch a larger ferry to Jersey.

Jean Preece and her husband, John Doswell, also part owners of the *Harvey*, were among the many others at Pier 63 helping out that day. The two of them, Jean's mother, and their dog headed "instinctively," as Doswell put it, toward Pier 63 and the waterfront when they heard the news. For them, Pier 63 was where the fireboat was; it was where they hung out: "Obviously you want to do something. The minute we headed over to the waterfront, we . . . realized the evacuation was happening. . . . It was just the obvious thing to do . . . to jump in and help with lines." Doswell retrieved his bullhorn from his boat, and Preece started directing people to boats heading to different locations. They used twine to make proper queues. As Krevey explained, "Kite string would have [had] the same effect as the strongest rope. . . . People want to be orderly . . . and it's hard without something physical to see."

There are many kinds of tasks a person can do to help out in a disaster that do not rely on one's principal occupational skill. Volunteers and emergent groups often just serve food, inventory supplies, answer phones, make deliveries, stack boxes—the kinds of routine tasks that people are familiar with from their everyday lives. People who share skills often share a vision of how to organize their helping activities, even when—as was the case at Pier 63—those activities have nothing to do with their specific skill sets. Krevey had never organized an evacuation before, and Preece was no expert in commuting queues. As we have seen from the history of trade guilds and unions, and in workplace politics over technological change, skills provide the common ground by which people come to see that they have similar interests and can perform similar functions (Grzyb 1990).

We heard that at the Liberty Landing Marina, another site to which evacuees were transported, a young paramedic who was looking at the sight unfolding across the river started organizing the decontamination area. It was nothing fancy—hoses and buckets—but he reportedly anticipated the condition of people who would be heading over. Eventually, this operation moved off the docks, where it was creating a backup in unloading people from subsequent boats, and up to the parking lot of the restaurant area where ambulances and local emergency vehicles began to arrive. Other spectators started bringing down chairs from the restaurant's upper deck so that the evacuees would have a place to sit and rest. A woman from Canada—some people thought her name was Adele—started doing triage and comforting people. A bartender came down on his bicycle and helped. Several of our interviewees recalled a large number of physicians or medical students coming over from a serendipitously located medical conference, anticipating the many injured who we now know would never come. Almost everyone we talked with recalled a sense of camaraderie that they could not quite explain. "Yeah, you knew you had to help," one person who was at the marina stated. "You knew you just couldn't sit there."

Bruce Boyle, one of the owners of the Liberty Landing Marina, recalled a woman who approached him with an offer to help coordinate the growing operations at Liberty Landing. She worked for a large company, Boyle remembered, and seemed to be quite senior by the way she took charge of the phones in the office and set up a telephone hub on site. This effort turned out to be crucial for the growing response operation, given the difficulty in getting through to anyone by phone or radio after the towers went down. Later that evening, with the help of the ad hoc infrastructure she had established, the restaurant became a sort of command center for several federal agencies.

We know that hundreds, perhaps thousands, of people stepped up to help out in the aftermath of the attacks on the World Trade Center. Each contributed in some way. Unlike some captains and crew who operated a ferry service—for whom one might argue the evacuation activities were an extension of their normal routines—evacuation was not normally within the purview of many other mariners who assisted on 9/11. Even so, the captains and crews interpreted the situation within the context of their identities as mariners. In this process of diffuse sensemaking, each made sense of the response and emergent needs in a similar way.

Consider how three participants in the evacuation, each with a different background, came to the decision that he had something to contribute:

> I think the Coast Guard put out a call for boats to respond for evacuation purposes. I didn't hear it myself. [I was] told that's—that's what they did. Whether they did or they didn't, I think everybody—all the maritime community was running, moving in that direction anyway. . . . Everybody[,] we had made a conscious decision and we said . . . there's something going on here way above and beyond . . . normal procedure. We're sending everything we got up there. (Jack Ackerman, harbor pilot, South Street Seaport Museum Oral History)

> So we get back and we see these thousands of people walking north. . . . Every Tuesday there's a jazz thing called CD101, and the *Horizon*, 600-passenger charter boat, was scheduled to be

> here at two o'clock. So, the first thing I did was call them on the telephone and we were talking and they decided to come anyway. I talked to them about possibly being involved in a relief effort to get some of the people out, because I saw all these thousands of people and there just didn't seem to be any other way off the island. . . . A number of my friends all ended up here simply because there were people wandering around who didn't know where to go, and this has always been sort of a sanctuary for people. So, they and myself started getting some stanchions and ropes out so we could make orderly lines, should we start getting people organized to go on the boats. (John Krevey, pier operator, South Street Seaport Oral History)

> I saw this smoke from the World Trade Center. I immediately went back to my shop, . . . grabbed my marine radio, life jacket, and headed straight for lower Jersey City to hop on a ferry. I knew there was a fire; I knew that we had to do evacuation—"we" meaning the ferry company. I feel responsible for building the docks; therefore I felt very responsible for getting the people out of there safely. I knew at one point when I saw the intensity of the fire that there was a possibility that that ferry terminal at the World Trade Center might not be open. . . . It is our busiest terminal moving around 20,000 passengers a day. I knew that I might need to communicate with ferryboats to bring boats in elsewhere. That is why I grabbed my radio, so I was in constant contact with the ferries no matter where I was on the shoreline. (Paul Amico, ironworker, South Street Seaport Oral History)

These respondents determined that an evacuation would be necessary because of the information available to them, but each interpreted that information through different identity categories. Even those who responded to requests made from within or outside their organizations (as opposed to being self-deployed) made sense of the steps they would need to take within the context of their identities. For example, Amico's actions were closely connected to his identity categories as an ironworker, boater, and member of the harbor community. Lots of people saw the same things, but Amico, an ironworker, saw and noted ironworker-related aspects. He, in turn, built his sense of the situation on these perceptions and acted accordingly. He stated:

> My concern was the waterfront. . . . There was nothing I could do upland. . . . I'm a water-based person. . . . I'm not the middle-of-the-river guy. That's the ferry captains. [He tapped his finger on the chart open in front of us.] This strip here from the land to the edge of the water—that's my forte. So that's where I went.

It may, at first, seem somewhat roundabout to assert that boat operators' identities were important in an operation involving boats. But surely other capacities, other identities, were also available to these participants. For example, Furber, the *Harvey* fireboat crewmember who was also trained as a welder, left the vessel at one point to cut bodies out of crushed vehicles. Thus he began an entirely different phase of involvement, shifting from one sensemaking milieu into another. Identity brings with it many elements of shared knowledge that can then be deployed as well as a shared ethos toward particular kinds of action. Having people with congruent or compatible identities distributed throughout the area increased the chances of action-oriented connections between them, especially if that identity included a shared knowledge base.

In times of crisis, people are creative and flexible; they may see additional opportunities to help. Jessica DuLong, for example, had a psychology degree from Stanford and had been recently laid off from a dot-com. Less than a year before September 11, she had started working with the crew of the *Harvey*, inspired after a day of volunteering. By fall 2001, she was an assistant engineer for the retired fireboat. Stuck in Brooklyn that morning, she made her way by boat to the *Harvey* the next day, after calling Ivory. Lenny piped in and told her to head to the Navy Yard, ask for Marine 6, and tell "anybody wearing a white helmet or a shield or a hat [in other words, a fire officer] . . . that you're the assistant engineer on Fireboat *Harvey* and you have to get to your boat." It helped to have someone with significant experience in the fire department on hand to help get others in.

But, she recalled, "At the time I felt like, I felt pretty ineffectual on the boat. . . . Once it was set, it was set, you know." But she soon noticed the chaos of the supplies coming in by boat. Some of the goods were needed, of course, but much of it was not. DuLong's role evolved: "I [knew] I wasn't qualified to be on the pile . . . but I knew I could make [the supplies coming in] more organized. . . . I knew there was a ton of need for that. . . . I'm seeing people not getting what they needed. . . . It just felt totally inefficient and I want efficiency."

DuLong's role as assistant engineer with the *Harvey* certainly influenced her starting point within the response, and she gravitated to helping out in this capacity. Yet, she was relatively new to the maritime community. When an opportunity presented itself for her to have what she perceived as a greater impact, she was able to envision a different role in the supply organization and temporarily move out of the maritime role. How does this move relate to identity? DuLong noted a desire for efficiency when she saw the chaos of the incoming supplies. Perhaps she drew something from her years honing organizational skills as a managing editor and in website content development. There is another plausible explanation: research shows that women are often relegated to disaster roles consistent with routine gender-role expectations (Alway, Belgrave, and Smith 1998; Enarson 2001; Fordham 1998), but this is often expressed in the context of men helping strangers while women help family. This is not to say that DuLong saw the need for, as she stated, "finding batteries and fold[ing] shirts to make it more organized" because of her gender, but it would not be inconsistent with prior research to assert that many of the men alongside her did not see the real need to organize the supplies as a priority because of *their* gender. We can speculate that this movement to another task would be something she might have been less likely to do had the event occurred a decade later, after she had spent many years working on the *Harvey*.

The influence of identity was not reserved for those with a mariner connection. Patrick Harris, the captain of the sailboat *Ventura* who joined the *Royal Princess* and whom we introduced in Chapter 1, recalled with great admiration the orderliness of people waiting for transport at Pier 63 and the role of a particular bartender in making those evacuees feel comfortable and at least a little more at ease:

> [The] bartender from the little snack bar on Pier 63 was walking up and down the line with nacho chips, and he had drizzled something on them, giving them to people, laughing, making them feel good . . . and people were taking handfuls of chips. And I remember throughout the day. . . . every time we came back in there . . . [he] would throw us a couple of bottles of water, and he kept passing chips out.

This bartender, too, found a familiar role. In popular culture, the bartender is the informal counselor, serving you while listening to your problems or stories and making you feel just a little more comfortable than when you walked in. It may seem on the surface that chatting with people waiting on a pier and offering them a little food or a bottle of water would not amount to much, but it was something this individual could identify as making a difference to those who were experiencing extremely stressful circumstances.

Recall Furber, who temporarily left his activities on the *Harvey* and employed his welding skills at Ground Zero. Being a mariner, after all, was not his only identity and not the only way he could help. Likewise, Peterson left his boat and became primarily involved in landward activities, citing his extensive training and experience in incident management. Many times mariners, including DuLong, Furber, and Peterson, gravitated toward helping out on their boats because that role was closely identified with what they saw as their master status. But sometimes other roles emerged as secondary options when the mariner role was not available or no longer demanded their attention.

Habits of Mind, Habits of Hand

Boat operators talked about how they or others just saw something that needed to be done and did it. Actions provide cues that others can decode and further act upon. One version of this phenomenon is *keynoting*, in which a strong cue provided in times of ambiguity helps define a situation for others (Turner and Killian 1987). Keynoting can be as explicit as a direct statement or as subtle as a gesture. In a fire, for example, people typically stand around and confer with one another on the nature of the crisis, a process known as *milling* (Turner and Killian 1987), until eventually someone goes for the door. At that point, others follow, even if that person—the keynoter—says nothing.

Weick's (1995) definition of sensemaking is social in that an individual's identity, skill, and knowledge—each of which is defined in relation to others—are the apparatuses for extracting cues from the environment. A shared identity, especially a shared skill-based identity, suggests shared knowledge. In the case of the waterborne evacuation of Manhattan, mariners' membership in a "community of practice" (Hutchins 1994: 79) provided an operational schema first for action and later for understanding. The principles of practice that they recognized were the foundations of an *epistemic network*—a thinking network—in which participants could anticipate needs of those whom they had not met (Rochlin 1989: 161).

Many of the boat operators and waterfront workers shared portions of the same knowledge base through their occupations as merchant mariners, harbor pilots, or Coast Guard personnel. Hence their training (Comfort 1999) provided a preexisting foundation for forming expectations regarding how the boats should be handled, how they should be loaded, the capabilities of different vessels, the application of maneuvering rules, and so on. A close corollary of shared knowledge, in this instance, is the development of a sense of "shared risk," which Louise Comfort (1999: 4) emphasizes is important in fostering the ability of disparate groups to work together to meet common needs. Shared risk, however, depends on a sense of shared values or shared assets that are threatened. In an individualistic culture, such as the United States, we do not normally think of ourselves as sharing assets, except those we perceive to hold in trust as citizens. As a consequence, a shared knowledge base would seem to be critical in sensemaking. Because the waterfront workers shared behavioral norms as well as aspects of the same professional skill sets, they were able to similarly interpret the emerging needs.

Mariners gather their skills—and thereby their acculturation—in various ways. There is no single path for training, certification, and licensure in the merchant marine. Some acquire sea service in the U.S. Navy, U.S. Coast Guard, or U.S. Army. The state maritime academies (in Massachusetts, Maine, New York, Texas, California, and Michigan) and the federal academy at Kings Point, New York, provide an intense program of training in a four-year baccalaureate program. These schools, which emphasize a military or quasi-military regimen, primarily offer degrees in marine transportation (or nautical science) or marine engineering, but some of them also offer majors in marine safety and environmental protection, ocean science, naval architecture, and others. At graduation, students (variously called cadets or midshipmen) are qualified to take the U.S. Coast Guard examinations for third mate or third assistant engineer.

But few of the 9/11 boat operators had trained at one of these academies. Most, instead, "came up through the hawse pipe." This metaphor refers to the hole in the deck through which the anchor chain runs, meaning that they worked their way up through the ranks—sort of pulling themselves up by nautical bootstraps.

Even with the diversity of training paths, it is possible to identify some habits or practices of mariners who might be regarded as successful. We are referring here not to technical competence—obviously an understanding of the use and limitations of equipment, of the functioning of a ship's systems, and of ship handling and seamanship are essential. Probably most important is the ability to maintain an attitude of vigilance for sustained periods. The phrase *attention to detail* recurs throughout a mariner's career, and it means simply to be always looking and listening: looking for the piece of equipment or cargo that needs to be tied down more securely, looking for possible sources of fire, or listening for changes in the sound of the ship or machinery. In general, continuous watchfulness and alertness for the unusual, the puzzling, and the anomalous are perhaps the most important habits of the seafarer. The mariner learns early in a career to trust no one particular thing.

The words of those we interviewed reflected this same expectation of vigilance, but in a way that reflects the effect it had on the positive interactions between various crew members. As DuLong from the *Harvey* described it, "Everything was so dire and so severe and so high stakes. . . . [You] get hypervigilant. And you are keyed up and keyed in." One mariner told us, "So much of our business is based on common sense and dealing with reality. I mean, it may not be what it should be, it may not be what you want it to be, but you deal with what you got." Attention to detail and constant checking are among the mariner's distinguishing behaviors, an institutional obsessive-compulsive disorder. In the marine world, doubt is a virtue, and mariners are trained to check and recheck. Are the mooring lines secure? Is a line fraying? Are any containers working loose? Is water leaking in anywhere? Is the GPS working right? Automated systems are to be trusted only provisionally; mariners prefer to see, touch, and feel what they are working with, and they know that if an automated system fails and they have not been running checks on it, the blame for any accident will fall to them.

A certain degree of independence and creativity is required as well. Some maritime accidents have occurred because mariners, perhaps fearing a reprimand, have failed to present the commanding officer with contradictory information (National Research Council 1994; Perrow 1984). In those situations, researchers point to the rigid hierarchy as fostering the potential for error. The concept of *bridge resource management*, patterned after aviation's cockpit resource management, has emerged to create a style of navigation practice in which information is gathered and shared by members of a team, checking and rechecking one another, rather than being gathered and processed in a linear, hierarchical fashion. Within this schema, an officer must be willing to exert his or her influence on the evolution of a situation, even if that means contradicting the captain.

To understand how pilots on September 11 so smoothly inhabited their new roles as disaster managers, we need to probe a little more deeply into what the pilot's job normally involves. Oceangoing ships were not involved in the boat evacuation, but the rules for their handling tell us something about the skills and adaptability of ship pilots. Consider a ship approaching a harbor from seaward. In New York Harbor, the pilot boat *New York* is stationed at the harbor entrance. It serves as a dispatch center for local pilots, who wait on board to be sent to incoming vessels whose crew lack knowledge of New York Harbor. When a ship arrives and requests a pilot, the pilot rides over in a small boat and climbs aboard the ship on a rope ladder hanging over the side. It can be a long climb. Here is the first assessment the pilot makes: the ladder had better be in good condition, because the pilot is stepping from a wobbling, bouncing boat onto the rungs. The maintenance and rigging of the ladders is such an important and potentially dangerous task that an international committee exists to monitor inspection methods.

Upon reaching the main deck, the pilot is met by one of the ship's officers and escorted to the bridge. What happens next is a ritual defined by a mixture of custom, technical necessity, and the need for everyone to develop shared expectations to coordinate the navigation of the vessel and to guard against future liability. The master—or the ship's captain—and pilot exchange introductions and discuss the ship's engines and other equipment. Here, the pilot makes another assessment, this time regarding the condition of the ship and the alertness and ability of the crew. Ships' crews have been downsized, with the remaining crew members handling more and more tasks in spite of the need for alert and vigilant officers—a need illustrated in accident reports. Crews are often exhausted, especially on ships that make frequent port calls. The pilot may in fact be the only truly alert person on the bridge as the ship enters the most challenging segment of its voyage.

If a pilot reporting aboard ship finds sleepy or disconnected crews, that experience becomes part of the pilot's own toolkit of vigilance and preparation. Although the pilot does not live with the ship, for the two or three hours on board, he or she lives with the legacy of decisions made by the ship's owner and regulators—on top of whatever is happening with the crew. No wonder pilots develop a sense of wariness; like the boat operators sailing for Manhattan, not knowing what to expect, the pilots who take ships in from sea also do not know what to expect.

One person we talked with described the pilots in this way: "You talk to pilots anywhere in the world [and] I think they're all going to say, 'Yeah, you got to be ready for anything. You got to be ready for whatever situation confronts you.' Whether it's a humdrum, ho-hum day or it's a, you know, kind of action-packed, scary, event-filled day." He went on:

> You could be aboard a container ship one day, an oil tanker the next day, a cruise ship . . . a navy ship . . . a tug and tow, heavy

> lift ship, all kinds of different ships. It's all a little bit different, but what goes through a pilot's head is, "What's different about this ship? . . . What do I need to know about the propulsion and the steering? . . . Where is it going? What am I going to expect on the way there?" You know, like, "What other vessels am I going to have to overtake or pass or things like that?" . . . Basically, [the pilots] get aboard and they plan. . . . They don't [just] deal with exactly what is exactly right in front of their face, but they also try to project and figure out what they're going to encounter on the way in. And they have to be ready for anything, because ships break down . . . when you least expect it. Weather changes can occur . . . relatively quickly. . . . I don't think there's anybody who can say they have never experienced an emergency situation. It happens. . . . So you have to be ready for anything. So that kind of helps in situations like this because . . . they're not supposed to get flustered easily. They're all human beings of course . . . [but you are supposed to] try to figure out what's important and deal with it.

Yet for all the skill and reliability of the pilot, and his or her vast knowledge and training, the ship's crew has a different set of responsibilities, and the ship's captain retains full responsibility for the ship. Pilots, in spite of their revered status in the industry, are not infallible. They make mistakes, as demonstrated in numerous examples found in accident-investigation reports. In the near-miss collision between the cruise ship *Statendam* and the tug-barge *Belleisle Sound*, the Canadian Transportation Safety Board (TSB 1998) found that in this accident, the crew did not monitor the pilot and did not realize he had become disoriented, which led to a near-miss with a tug-barge unit. The *Statendam*'s necessary avoidance maneuver was so severe that it sloshed water from a swimming pool into a passageway. In the grounding of the *Raven Arrow*, the pilot lost track of where he was, misidentified his turning cues, and turned early for an upcoming course change (TSB 1999). Again, the ship's officers had not been monitoring what was supposed to happen. Thus, the captain and crew need to be ready to correct a pilot's error, even to relieve him or her of the *conn*, or the responsibility of ordering the course and speed of the ship. The pilot, in other words, is never the ship's master. These sorts of accidents almost never happen, but sustained vigilance is necessary, because going to sea requires preparing for events that almost never happen.

Suspicion essentially becomes part of the principle of good seamanship; in the demanding marine environment, one is well advised to trust nothing, and thus suspicion and distrust become part of the seafarer's risk-management equipment. Mariners are acutely aware of their somewhat-conflicted status within the industry. They know that primary responsibility for the safe operation of ships rests with them, yet they also know that they are regarded as expendable components of the system. Safety, for instance, can be seen as a drag on operating expenses rather than a contributor to profits. Mariners not only understand that they are expendable; they also sense that new technology has far more to do with thinning their ranks than with reducing risks.

The wise mariner develops habits of looking, thinking, and questioning, and exhibits the qualities of vigilance, skepticism, and doubt at all times. Their professional lives are always on the edge of crisis, looking ahead to possible dangers, planning maneuvers far in advance, having an escape strategy for a dicey traffic situation, having a little extra speed in reserve, just in case. These habits, it turns out, are excellent preparation for disaster response.

The Importance of Strong Communities

From popular descriptions of the response to 9/11, one might be tempted to describe New York's maritime community as "close-knit." Based on these descriptions, one might be tempted to conclude that a successful unscripted disaster response relies on a preexisting, cohesive community. But while community is definitely important, it is not necessarily so in the ways that one might expect. New York's maritime companies were, and are, competitors as well as colleagues. Even so, it is definitely the case that these mariners *knew* one another, and therefore they *knew* how the others thought.

Our interview subjects stressed this point—that "everyone knew each other"—repeatedly. This is more likely true at the level of company officials and boat captains, maybe less so among the crew, but it emerges strongly as a factor in easing the coordination among the different boats. Just as importantly, especially in a national security situation, it gave the participants credibility. That level of prior acquaintance came simply from many years of working on the same waterway, of being conspicuous on harbor planning committees, of being part of trade organizations, and of providing services to one another. Miller Launch, for example, is a company that specializes in shuttling supplies and people from anchored ships to and from the waterfront. It is known to all the pilots and boat operators in the harbor. Someone said, "There's only, like, one degree of separation in the maritime industry, and by that I mean if I didn't know someone personally, the pilots knew them." Towing companies, such as McAllister Brothers, with its 150-year history, are renowned in marine circles. And who has not heard of the famous Circle Line sightseeing boats? All the employees of these companies see one another's boats and hear the captains, mates, company officers, and dispatchers talk to one another on the radio all the time.

Part of a company's business competence is knowing other operators' capabilities. The competitive margins are slim, however—freight rates can be only so low, for example. Competition means knowing one another's equipment and general expertise. Moreover, mariners often shift jobs from one company to another:

> We're all competitors, and we're all customers. That is, we sell at any time to almost any of those companies, and, vice versa, we'll buy their boats. So . . . we all know each other, and even though we compete for business, there's . . . that customer-competitor thing that always goes on. So generally speaking, we know people, we know everybody, and to some extent, there's a movement of people between companies.

These acquaintances were the carrier wave for organizing something larger out of the *individual* experience of mariners seeing and thinking the same thing. Individuals came together based on familiarity, shared skills and habits of mind, and a shared knowledge of the geography of New York and New Jersey waterways. That familiarity extended across the different companies and across public and private sectors—the Coast Guard was part of that network.

Of course, just knowing people does not make you organized, and just getting organized does not mean you will stay that way in a crisis. Weick (1993) shows that in his study of the Mann Gulch case. Collapsing organizations are common in industrial calamities as people begin to lose track of the status of their technology, become confused, stop talking with one another, create conditions for harmful interactions, and act without thinking (Perrow 1984; Weick 1990, 1993). In disaster management, the move toward reliance on formal, centralized systems (which we discuss further in Chapter 5) is intended to prevent organizational collapse, but the assumption that a disaster is a "thing" to be managed rather than a series of events to be understood can also create conditions that lead to performance failures. Because the boat operators on 9/11 fragmented their work, they could keep track of their technology—boats and systems for traffic management; they created conditions that minimized the need for not only discussion among boat operators but also interactions with subsystems.

We have talked a lot about *community* in this chapter, because we observed evidence of it and because the mariners themselves specifically mentioned it so often. But in understanding disaster as an event that affects a community—and in understanding what a community-based response is—we have to examine the term more closely. Communities are made of individuals, groups, factions, and interests competing for money, resources, and publicity. Communities are not things, but *sociopolitical ecologies* (Peacock and Ragsdale 1997) made up of people with very different perspectives on many issues. Those working on the harbor and even living in the marinas described the positive sense of community, even if we can imagine there must have been occasional disputes or disagreements.

As one of them said:

> It's a whole community in there. I mean, you live in Manhattan, you don't know who your neighbor is for twenty years, and in the marina everybody knows everybody, everybody watch-

> es out for everybody, everybody helps everybody. I mean, if you're sitting there having a beer and your buddy's coming in, or even if he's not your buddy, it's just somebody with a boat, you get out and take a line to just help him out in case they're having trouble, because you know that might happen to you sometime.

The ferryboat operators had an additional membership in a different community: commuters. Many felt a deep concern for the passengers they carried back and forth to and from Manhattan every day. These were people with whom they shared greetings, handshakes, and stories. The ferryboats were providing a service, and, as with most service industries, the crews identified with a cast of characters who made up their daily routines. When the planes struck and the towers fell, the boat operators were worried about the fates of these people they had come to know. When they did not see them for days and weeks, they looked for them in the obituary pages of local newspapers.

Harbor pilots and Coast Guard officers not only have access to different spheres of local knowledge; they also have different relationships to the community. Commercial mariners appreciate the safety and rescue functions of the Coast Guard, yet they are also wary of the Coast Guard's policing and disciplinary powers. And yet, during the waterborne evacuation, Coast Guard officers self-identified as part of the broader harbor community. They could have tried to do all the work themselves, but they did not. Instead, they blended in, working closely with harbor pilots aboard the pilots' boat. They saw themselves as part of the community rather than separate from it.

One harbor pilot described the sense of community and camaraderie across the harbor, saying, "We may not know their names. We may not be able to put faces with voices, but we talk to these people every single day." Such events as Fleet Week and Operation Sail, as well as meetings of the Harbor Operations Committee, encourage a great deal of interaction. The conversations might be about port improvements or dredging, but they bring people to the table. As one harbor pilot put it, "Those connections became valuable." One mariner emphasized that no one boat or person should get credit for what worked. Rather, the evacuation must be considered an effort by the entire harbor community:

> The working harbor. It really saved the day. I mean, can you imagine? Where were those people going to go if there weren't boats to get them off the island? . . . I just never saw so many people walking over to the waterfront, and I don't know what would have happened to them.

On Thursday, September 14, when a rumor spread of another building collapsing, people hurried to nearby boats and headed back over to New Jersey. One mariner recalled a man who would not get off the boat when they reached the Jersey shore, insisting on immediately returning to Manhattan. He introduced himself as a steelworker and stated that he needed to get back to the site to work.

Identity, skills, community, an ethos of rescue—we can identify these elements as fundamental to provoking the boat operations on 9/11. The waterfront workers, and others connected with the life of the harbor, interpreted what they saw based on who they were, their experiences, and what skills or resources they had available. Some shifted in and out of roles, trading one disaster job for another, reading the environment and matching what they saw to what they knew—or to what they did not. Amico did not put himself in the middle of the Hudson; he reserved that place for boat operators. He found his niche at the water's edge.

We are often asked whether all this was not just natural—after all, even some of the participants talked of their "instincts." Here, we confront one of the paradoxes of disaster. In some senses, what a lot of these people did was indeed automatic. They wanted to do something. They wanted to help. But others were more skeptical, and instinct was not enough. It was modulated and filtered through contact with others, people confirming or reinforcing their interpretations and choices. Often, there was a strong component of deliberation. As urgent as the event was, there was still some time to consider what to do. The staff at Reinauer talked through their decision. Phillips, Miano, and Haywood figured out what to do together, in their little group, and pulled other people in. Some mariners held their boats near Manhattan, just raring to go, but first asked permission from the Coast Guard. There was conversation, thinking, weighing of options. Hepburn started alone but went to Pier 63, which Krevey described as a "sanctuary" of sorts. It is where the *Harvey* was normally docked. People hung out there. On 9/11, the members of this community, as varied as it might have been, blended the pieces of their ordinary lives to assemble an emergent disaster response. And within that emergent response effort, they were able to start taking action and helping others take action as well.

3 /

Making Sense and Taking Action

Liberty Landing Marina, a private marina in Jersey City, New Jersey, about a mile west-southwest of North Cove Marina on Manhattan, had several hundred tenants in September 2001. Some of the slips were occupied by boats belonging to people who traveled a great distance and were there only temporarily. Others belonged to regulars who lived on their boats or stayed there when it was easier than heading home for the night. The marina also had a water taxi, *The Little Lady*, that served commuters who worked just across the river in Lower Manhattan, a marina storage area, a restaurant, and commercial tenants.

Adam Tierney—general manager for Liberty Landing Marina for the past five years—had just arrived at work when the first plane hit, as had dockmaster Dan Chuli. Chuli was talking to one of the boat owners when they heard a low bang and started to see the smoke from World Trade Center 1. Thinking it was a gas explosion, Chuli ran and got his camera. P.J. Campbell, a deckhand with *The Little Lady*, had the morning off and had been sleeping in. He lived at the marina, and he, too, went down to the gangway with a camera after the first impact. That was when the second hijacked plane flew overhead. "I thought I was still dreaming," he said. Meanwhile, along with his crew, Captain Keith Nevrincean had been ferrying people on *The Little Lady* since 5:30 or 6:00 A.M. Nevrincean and his boat were at the base of the World Trade Center, in North Cove Marina, when the first plane struck. As smoke from the explosion covered the sky, glass and other debris showered down on them. Thinking it was an accident, they headed back to New Jersey and began loading up for another run. The second plane hit as they were en route back to Manhattan. "Then we knew it was terrorism," he said. "[It] was so surreal." Bruce Boyle, part owner of the marina, waved the boat back in.

Bob Campbell, a friend of P.J.'s [no relation], was still on his boat that morning when the first plane struck, late for work since the Giants game he had been watching had run late the night before. He headed to the slip and saw port captain Bill Davis. "Captain Billy," he said, "we got to go." They went up to the Lightship Bar and Grille and saw a few state troopers.

P.J. Campbell was still looking at the burning towers and remembered saying to someone beside him, "Well, the architects must be pretty impressed by the way. . . ." He was going to say, "They were holding up," but he did not get to finish his sentence; World Trade Center 2 began to come down. Nearly simultaneously, someone urged Captain Davis to go and help, and Davis heard the call from the Coast Guard for all available boats. Davis, Nevrincean, P.J. Campbell, Bob Campbell, and Jerry Martin (another tenant) formed a makeshift crew for *The Little Lady*. They untied the boat and were across the river in a few minutes. They shut the windows and ventilation and went in by radar through the plume of debris. They were not alone in relying on radar: one captain coming from Highlands said it was like they "were going into Mount St. Helens."

Even with the radar, *The Little Lady* hit the seawall at North Cove. The seawall was capped with an iron railing, and they found that they were a good twelve feet below it due to low tide. Still, people desperate to get off the island started climbing over the rails. They tried to back the boat up to another spot, but when that did not work, they headed back into North Cove, where the visibility was getting better. Davis was "hitting the horn on the loud speaker, yelling, 'Come this way, follow our voices,' that kind of thing." The Campbell pair could not quite recall when the second tower went down, although P.J. thought they were heading into it, near North Cove. Davis, on the other hand, recalled they were heading into South Cove. One might think that these key moments would be forever etched in participants' minds, but it is often the case that responders' memories disagree on this sort of detail. So much was going on in such a short time period, and, just as importantly, the environment was disorienting.

The first people trying to board the vessel were covered in dust, and many had minor injuries. The crew of *The Little Lady* recalled one man vomiting dust, all the while apologizing for doing so on the deck. A guy assisting him stayed on board the boat to help people fleeing the city even after they unloaded the other passengers at Liberty Landing. "Then he just disappeared," said P.J. Campbell. This behavior would turn out to be quite typical: someone, unknown to others, would pitch in for a while and then move on. Perhaps these people headed home, or perhaps they went to help in some other way at some other place.

The Little Lady took many trips back and forth, starting from North Cove and later from South Cove. Captain Davis checked in every hour with the Coast Guard to report the number of passengers it was carrying and where it was embarking and disembarking passengers. When the boat returned to the marina, tenants who owned private boats began throwing their life vests on board to add to *The Little Lady*'s supply. Among the crew, everything went smoothly. As one of them described it, "We all just got in the game, leaned into it, and used our expertise as best we could without any kind of personality conflicts." Another crewmember stated, "Everybody realized that everybody worked together like they'd been practicing it for a thousand years." None of them recalled receiving any clear instructions from anyone on the VHF radio. "Every single VHF channel was flooded," one of them said. Instead, they heard a lot of misinformation. They heard of attacks in Philadelphia and Trenton, which, of course, never took place. They heard that more than one hundred injured children were waiting to be evacuated from South Cove. Fortunately, this also turned out to be just another rumor.

This is the norm in the early stages of disaster—little information or ambiguous information. But people start to piece together what happened. They may not know the full extent of the calamity, but they start to work in their own local environments. From the flooded streets of New Orleans after Hurricane Katrina to the tsunami-swept coastline of India and Sri Lanka in 2004, official and unofficial first responders often are not quite sure where they are, where they are going, or what is going on.

Via sensemaking, according to Weick (1995: 30), people create environments that they can understand. In this view, there is no fixed environment in which people act; rather, their actions continually shape or build the environment in which they find themselves, which is in turn something that needs to be made sense of. Sensemaking is "social" (38) in that "conduct is contingent on the conduct of others, whether those others are imagined or physically present" (39). In other words, to the extent that sensemaking requires action, those actions are affected by others' actions as well. Individual thought, moreover, is understood to be affected by social contact with other people and organizations (5).

All of us possess some knowledge of our local environments that we may find difficult to put into words. This deep knowledge of our familiar surroundings forms part of what social scientists refer to as "tacit knowledge," a sort of background knowledge that is more absorbed and theoretically transmitted than taught (Wood 1987: 3). But while this phenomenon is, to a certain extent, universal, mariners have access to specific tools and skills that make them particularly well positioned to "chart unfamiliar waters," as the saying goes. During the waterborne evacuation of Manhattan, those navigational skills combined with a deep knowledge of the local environment to make action seem not only necessary but also possible. What was needed next was some exploration to see how the environment had changed and where their skills would fit in.

The Unfamiliar Environment

The people we spoke with knew Manhattan well, but on that day and over the next few days, the environment in which they found themselves was anything but familiar. As one person we interviewed recalled:

> Nothing looked like it did [before]. Everything was covered in dust, and the sky was actually . . . it was a bright sunny day, but it was dark up there. It wasn't pitch black, but it was gloomy because of the smoke covering and cutting out the sun.

As it stands, many of the other mariners we spoke with had similar memories of the environment. One person described New York City this way:

> Dead quiet, and it looked like you just had a snowstorm, except for it was gray. Everything was gray, covered in dirt—not dirt, but you know what I mean. And papers flying all over the place. There was a lot of shoes in that water for some reason. I remember that. I don't know how they got there.

Another mariner who was on a tugboat that day described it in a similar way:

> There was no color in Manhattan that day, everyone was gray. There was no black, white, yellow, red, brown. Eyeballs and teeth was all you saw. . . . They looked at you; you just saw their eyeballs and their teeth.

This person described the aftermath of the towers' collapse:

> It was . . . in contrast to how television reported it—which was constantly the planes hitting [the towers]. . . . When you were down there [later that day], it was this murky silence. . . . Everything down there was slow motion. And all covered with white. Totally opposite [of] how other people had seen it [on television].

Those who entered the disaster area were the explorers of a changed earth. Familiar landmarks were gone or obscured, yet other aspects of the harbor, procedural, and organizational environments were still familiar, as least to some of the observers. It is a strange contradiction: in many ways, we heard how everything seemed different or foreign, but people were also able to find familiar touchstones that affected the direction of their work. These touchstones served as—sometimes precarious—starting points. But time was of the essence. Consequently, those responders exploring the disaster environment needed to quickly adjust their senses of it, simultaneously acting and completing tasks while still trying to figure out what that altered landscape meant in terms of necessary action: thinking while doing, or sensemaking while acting. One of the ways those we spoke with were able to engage in this process so efficiently was to share their newfound interpretations of the environment with others in various ways.

Active Sharing of Knowledge

As a crisis develops, not everyone who is involved or who will be has all the required information or the same situational awareness to comprehend fully the ongoing events, needs, and actions. Yet shared knowledge is important for modulated action. Our conversations with participants in the waterborne evacuation identified numerous instances of *sense lending*: instances in which information and knowledge passed between individuals and across organizations. Thus, we interpret the "shared" in shared knowledge as an active process of sharing, a sharing that helps enact the environment.

A number of those who were involved in this event were local harbor pilots. The pilots who participated were, indeed, local experts with an encyclopedic knowledge of the harbor developed over years of apprenticeship and professional service. For example, their jobs took them across all of New York Harbor, and they had to be familiar with a variety of vessel types. Pilots' knowledge also differs from that of the Coast Guard in that they spend their entire professional lives in one place. Because the ship-to-ship radios still worked, the pilots' local knowledge became available to support the operation (at least for those who were listening to their radios), and the pilot boat *New York* became a kind of traffic control center with both harbor pilots and Coast Guard personnel aboard.

A Coast Guard officer described the knowledge that personnel within that organization have about the harbor:

> I think the thing that prepared us for this evolution [or event] is just how often and how much we do in New York Harbor. We're so familiar with it. Almost every day we're going out. We know the area, we know landmarks, we've worked with many of the pilots and boats and tugs and run into them every day. And we generally go out and do a hard day's physical labor. . . . We're constantly adapting to whatever the situation is and this was just another adaptation to a particular situation. (Steven Whitrack, Coast Guard buoy tender captain, South Street Seaport Oral History)

But Coast Guard personnel are transferred to new duty stations every few years. A second statement, offered by a harbor pilot, contrasts the Coast Guard's perspective on the kind of knowledge that pilots had:

> One of the advantages that the maritime industry people have over the Coast Guard is that we know most of the vessels that work in the harbor because we've been here for our life. We kind of know the size of the vessels, we know the capabilities of the vessels, we can understand where they can fit and where they can't fit. So having the pilots out there was really a distinct advantage for this particular instance. We were able to deploy these people in a pretty efficient fashion. It worked out well. You knew . . . the height of the tug, how high the deck was, what wall he could get alongside, and people could step into the boat rather than . . . have to climb down or climb up into a boat. That worked out well. (Jack Ackerman, harbor pilot, South Street Seaport Oral History)

Certainly the Coast Guard could have taken over the operation, at least in principle; they still had the authority and the responsibility. But they also needed the maritime resources and the extent of connections that comprised the commercial maritime community. In the interests of sharing information, the pilots and the Coast Guard worked more closely together than usual. They explicitly shared information, by sharing a radio band and by sharing the physical space of the pilot's boat.

Sometimes communication took more mundane forms. When the radio became too busy, for example, volunteers and first responders sometimes reverted to hand signals. For instance, ferryboat operator Rich Naruszewicz told those on shore he would give them a sign as he approached to indicate the situation on board: "My right hand said how many ambulances we need[ed], and my left hand would say how many people were in real bad shape." Or they just hung handwritten or spraypainted signs for ferry destinations: straightforward written communication. In this way, evacuees could see which boat might take them at least close to where they lived and make it easier for them to get a ride home with a friend or relative.

Others emphasized the importance, quite literally, of being heard. John Krevey, who organized the evacuees at Pier 63, told us:

> Everybody [in the crowd] had the same question. They wanted to go to particular places in New Jersey; they wanted to go to Hoboken. . . . I would never answer [an individual's] questions; I'd always get on the megaphone because I could answer a thousand people who also always had the same questions. (John Krevey, pier operator, South Street Seaport Oral History)

Years later, Krevey reinforced this statement:

> The single most important piece of equipment which I own—and which I own three of—that prepared us for this emergency were three megaphones, three battery-operated power-amplified megaphones. Made the difference between giving us the authority to control the situation exclusively above anybody else. I don't care what their badge or their uniform [was]. . . . They could have been the chief of police. [We] had more authority than anybody in the world, simply because we had the power of the volume of the voice of the megaphone.

As we have seen, actions, too, can convey information. In the previous chapter, we mentioned Paul Amico, an ironworker who did a lot of work for New York Waterway. When Amico made it to Manhattan, he saw that iron railings blocked passengers' paths. Amico made a split-second decision to cut the fences down, grabbing some oxygen acetylene torches from his kayak club's boathouse and immediately getting to work. Amico did not necessarily have access to all the information that the boat operators, pilots, and Coast Guard had, but he was able to make sense of the evolving situation from his geographic knowledge of the harbor and the waterfront infrastructure. Moreover, through his actions, he *lent his sense* to the other participants via the change his actions inscribed into the terrain. His improvisation used the physical environment as a medium for "integrating information" (Comfort 1999: 25), and he made sense of the environment for the evacuees and boat operators in their new surroundings.

And here is something interesting: *fences were cut down wholesale along the waterfront*, many blocks from where Amico made his first cut. Just as the boat operators engaged in diffuse sensemaking, so too did those who participated in the cutting down of fences. Recall that the term *diffuse* here is used in a similar manner to *scattered*, where individuals are in distinct physical and social spaces but *make sense* of the response and emergent needs in similar ways. Here are just a few accounts of fence cutting:

> We nosed in at the New York Waterway landing site there, but it was kind of shallow water, it would draw ten feet, so we were starting to hit bottom, we couldn't get in there. We had to back out and go maybe fifty yards south to another pier which is an all wooden pier which is probably condemned. There was nobody on it; there wasn't any cleats or anything to tie to . . . just a lot of broken wood, and at the land end of the pier there was a chain-link fence and another wooden fence blocking anybody's entrance to the pier, you know because of safety reasons. . . . We had a lot of people on the pier trying to get everybody off in one location so we could get out of there, but I realized there was no way off the pier . . . , so we took some forcible entry tools. I had a gas powered demolition saw, I went up and cut the chain-link fence, a section of it down, pulled that out of the way, went into the next fence, cut this wooden

> fence down, broke it down with sledgehammers and axes to make a big hole so the ambulance crews can get through, the police crews can get through, so we get this big hole open in the fence, get everybody through. (Tom Sullivan, firefighter, South Street Seaport Oral History)

> And then as people were coming over [the seawall] there was kind of an ornamental, maybe a three-foot fence along the whole way. And as more and more people came we watched them climbing over that fence and trying to get on. And we had some Park Service folks onboard up on the bridge at the time. After watching about ten or twenty people struggling getting over this fence nearly falling and I said, "Can we cut it?" And she looked at me and said, "I don't think anybody would care right now." So we got out our torch and cut out a section out of the fence. And so we had our brow right on from our buoy deck up to this cut out section on the west Battery Park wall and it made it nice for people to be able to come right onboard easily. (Lieutenant Steven Whitrack, Coast Guard officer in command of a buoy tender, South Street Seaport Oral History)

Cutting away that section of fence, combined with the maneuvering capability of his vessel, allowed Steven Whitrack to make a stable platform to bring passengers aboard. Both Tom Sullivan and Whitrack, like Amico, enacted changes in the environment that subsequent evacuees could take advantage of.

Amico's experience demonstrates how the physical environment serves as a tablet on which *scattered* actors can figuratively write out parts of a story to be read, interpreted, and added to by other actors. Not everyone thought of cutting down fences, yet the act shaped the ways in which the evacuation and norms associated with it would emerge. His sensemaking enabled others to make sense, as boat operators understood where they could safely line up their ferries and evacuees saw where to go. And yet individuals may not recognize that they are receiving information, because they register it as an existing feature of the environment rather than as transmitted data.

Knowledge and Limits of Knowledge

Common knowledge provided the participants with an initial common environment of water, boats, and docks. Local geographic knowledge helped shape the response. People went to physical places that seemed reasonable from where a mariner might begin contributing. Knowing where people might tie up, knowing where people might disembark in safety: these and other reference points keyed different personnel to particular locations, depending on what they hoped to do. We see this at the individual level—for instance, individual boat operators who decided to assist—as well as at the coordinating level eventually assumed by the Coast Guard and harbor pilots.

In this case, many of the participants shared geographic knowledge—cartographic knowledge written down in the form of charts and navigational publications as well as remembered as accumulated operational lore. Norms in interaction, expected conduct, and navigation rules provided direction for action, but these norms were also overturned as needed to preserve a satisfactory operational environment.

In creating a shared environment, common knowledge is important. Louise K. Comfort (1999: 23) notes that "among professional emergency responders, this common knowledge base is achieved in large measure through training and experience." As Naruszewicz put it:

> I've been working in New York Harbor since I've been seventeen, and I scoured the waterfront. I studied the charts, what you can do, what they do here, what they don't do here. What kind of boat can go in here? What kind of a boat can't go in there? Being that I worked on a tugboat, we went to every pier there was.

Naruszewicz was a hawse-piper, having worked his way up to becoming an officer. He would go in and fill in for someone on a boat who was absent from work that day: "I can go in and learn a different way, you know, a different aspect to the way they did everything." The fact that so many of the boat operators, like Naruszewicz, learned their skills directly in New York Harbor meant that their cartographic knowledge ran deep.

Harbor pilots are the local navigational experts. During a fourteen-year training program, maneuvering thousands of ships around the harbor, they develop a comprehensive familiarity with the waterway. Part of a pilot's licensing exam is to draw a chart, with buoys, lights, depths, and landmarks, from memory. They know which boats can tie up at certain locations, considering the depth of water. They know tides, currents, hazards, and the location of submerged infrastructure, such as pipelines. That knowledge was made available to the Coast Guard officers who were working with the pilots, to the boat operators, and to people on shore who marshaled evacuees at three different points—on the southeast, south, and southwest reaches around the tip of Manhattan.

But, of course, cartographic knowledge was not the only kind of knowledge that mariners held. By virtue of their participation in waterway traffic on a regular basis, the participants had knowledge about the names of vessels, the other companies, and the people who ran them. This knowledge allowed members of the harbor community to see opportunities for action, ways of assembling these pieces of information in new ways. Consider, for instance, the ways in which the various operators recognized that some boats were more suited for certain purposes than others. For example, one of the big Army Corps of Engineers boats would have been ideal for bringing in heavy equipment, but it was not agile enough to get in and out from the docks quickly. *The Little Lady* was quick, with an upper and a lower deck (and therefore plenty of room for passengers). According to some of the mariners we interviewed, square bow vessels could make a landing anywhere and take passengers. Vessels with pointed prows—for instance, tugboats—were somewhat limited, as there was nothing for them to tie up to. They needed the sturdiness of the seawall to push their bows against as they held position. In other situations, wakes were rolling in, and little boats were vulnerable to slipping under pilings and walls. In other words, some boats were useful for certain tasks at certain places, while others were not. The local and professional knowledge of the mariners helped them determine those distinctions.

Tugboats certainly aided in the evacuation, and such vessels as Moran Towing's *Kathleen Turecamo* played important roles in this effort. Yet the tugs could not move as many people at a time as some of the large ferries. But, again, the participants realized this limitation. Someone on the waterfront quickly recognized, for example, that the *Rachel Marie*, the tug that Captain Robert Henry was piloting, would be more effective in another capacity. Soon Henry was charged with moving doctors and nurses from where they were arriving (along the Battery) to a location closer to St. Vincent's Hospital, in the West Village. The tug was ideally suited to move right up along the seawall in relatively shallow water. "These boats were made to bounce off the bottom," Henry explained. From there, the medical personnel could cross over the West Side Highway and quickly arrive at the hospital. The tug took three or four trips, carrying about six to eight people at a time. In other words, different vessels brought with them different capacities. It was a matter of determining who could do what and helping them find their niches.

Other participants in the evacuation with different occupational skill sets were able to anticipate certain exigencies that might confront boat operators who were focused on maneuvering their vessels and embarking passengers. A writer who reported on these events for a mariners' magazine observed:

> Among the maritime community's members, no potential problem seems to have been overlooked as individuals came forward to use their skills and company assets. Kurt Erlandson . . . anticipated the possibility of lines fouling propellers with so many boats operating in close quarters. One . . . dive crew had been working in the anchorage south of the Verrazano Narrows Bridge when the first plane hit the WTC. "I pulled them out of there and dispatched them to assist the evacuation. Then the rest of the crew and I arrived in New York . . . with a complete diving spread about 11:30, after we got clearance to go to the Battery," said Erlandson. . . . [T]he divers averaged six jobs a day clearing cables and hawsers from the response boats and tugboats. In their free time, they assisted coordinating the supply operations. (Aichele 2002)

Comfort (1999) has emphasized the importance of a shared vision among those who would collectively respond to a crisis situation. Our interpretation of this important concept, however, is that a shared vision does not exist as something that participants discover or that they "have" but rather as something that is created and potentially uneven among the various parties involved. The sharing is active, negotiated, and sometimes unknown even to the viewer. In the diving example, there was an overarching shared vision, a broadly definable goal, shared by the divers and the boat operators—to evacuate Manhattan—but not everyone shared the vision equally. The divers thus anticipated that boat operators might not look ahead to the possibility of tangled lines and ask for someone to stand by. Divers and boat operators were separate, independent, and differentiated. The divers' knowledge enabled looking ahead, enabling integration, enabling (diffuse) sensemaking, enabling action.

All of us have access to certain kinds of knowledge that we may find difficult to put into words, or different operating assumptions that guide our actions. The pilots understood immediately that bringing in boats with pointed prows would require a set of docking maneuvers that were different from those for boats with square bows; divers also understood immediately that that much boat traffic would inevitably create tangled lines. Together, these unarticulated skills and relationships to tools constituted a form of tacit knowledge.

Recall the statement by Jerry Grandinetti, one of the captains with VIP Cruises introduced in Chapter 1, who said that he was getting ready to do something but did not know what it would be. Similar statements were repeated over and over in interviews. In social science work, this phenomenon is called an *in vivo* code—something uttered by research subjects in their own language, their own vernacular, that can become analytically significant because it suggests that all of them are thinking the same way about something. That tells the scientist that some underlying phenomenon or shared experience among the subjects is worth thinking about some more. Listen to what they said:

> Kevin Tone: "We didn't know what we were gonna do, but we knew that, you know, obviously the impacted site [is] Manhattan. You know, the idea was to sail over there."

Ken Peterson: "Hey, we don't know what we're going [to] see, we don't know what we're going do, but we're going to go."

Ed Keegan: "We didn't know what we were going to do, but we had to go over there and do something."

Kerry Pearce: "We still didn't know, like, the magnitude of what had occurred, you know, we were just gonna hop on the boat to help the crews."

Richard Schoenlank: "We didn't know exactly what was going to happen . . . but we figured, we might need to evacuate ships."

Jim Sweeney: "We didn't know what we were going to see, and when you get reports, you don't really know when you're leaving here what really is happening."

At that point in the crisis, these boat operators had full creative freedom. They could provide first aid, shuttle people, transport equipment. But they were also thoughtful and watchful. Having no preconceived plans or expectations of what they would see, they could make sense afresh, avoiding the fallacy of anticipation in which they would be modulating their thinking to match an imagined state that did not exist. The boat operators were watchful and vigilant. *In having no idea what to do, they had lots of ideas of what to do.*

Consider, for instance, the experience of Jeff Wollman. Wollman worked for Reinauer Transportation as a construction manager in charge of building along the waterfront. Soon after the first collapse, Wollman and his son rode over to Manhattan on one of the Reinauer boats. They jumped ashore and walked inland a little way, meeting exhausted people as they went and directing them to the waterfront.

We can see more about the importance of identity in shaping this event by looking at someone who was not a mariner. His identity became a starting point for organizing activities in an unfamiliar environment. As a construction manager, Wollman was, of course, familiar with the use of heavy machinery—but his world of heavy machinery differs from that of sailors. For them, the boat is purpose and profession. For Wollman, a boat is a means of expedient transport to get him where he needs to be to build something. The boat is part of his supply chain of materiel and people. It is a platform he harnesses for lifting and digging, for moving the earth, dirt, cement, and steel that he uses to remold the landscape.

Wollman stated his skills for us matter-of-factly:

> I know diesel engines forever and again, I've been in the construction field since I was in high school. . . . I can drive almost anything. I drive my big cranes. I have big truck cranes that are kind of like an eighteen-wheeler. I've been driving heavy equipment and stuff all my life. I'm a crane operator and bulldozer operator and basically just from doing it all my life from being a kid.

He can hot-wire a truck, too. He is the kind of guy whom, if the world ends, you will want around to help rebuild it.

Construction managers, too, often face uncertainty and unfamiliar environments. They are used to sizing up situations; they are used to interpreting their environments and figuring out the connections of people, environment, and equipment for getting the job done and for finding new reference points. After Wollman and his son abandoned the evacuation operation, figuring it was best left in the hands of actual boat captains, they suddenly found themselves at a Burger King.

Wollman and his son looked around, taking in the scene of dust and devastation. They soon noticed others running around: firefighters, police officers, and even "a guy trying to untangle extension cords because they were trying to get power and there was power in certain buildings and other ones not." They started helping the cord untangler. Next they moved on to help others turn the Burger King into a command center: "We started clearing out—it was a two-floor Burger King, and on the second floor we started emptying chairs and moving things out of the way. We were going to make this like triage and start getting it set up, and we were, again, giving them a hand moving things around, getting things set up." But they found the restaurant too small. "The next thing, somebody came across [and] said, 'No, we're going to do the building across the street.' We went over, we started helping them clear and moving things out of the way and shoveling the street to try to get the debris out of the way." Wollman explained what he did next:

> Directly across from the World Trade [Center], there was a tractor trailer parked. [We walked over, saw] the cops by it, and [started] talking to them, and they said, "We got to get this thing moved, but we can't get it started." I am pretty handy with mechanics, so I said, "Pop the hood. Let me see if I can jump this thing," which I could. So, we got it up, we're starting to dig around in it, and then there was a cop in the seat, and he said, "I found the keys." The keys were under the seat, so he fires the thing up, and I'm sitting there with everybody else, and he's getting it running. I say, "Can you drive?" He goes, "Yeah, I'm good." [But he] couldn't get it going, so he says, "I can't drive this thing." I say, "You want me to do it?" He says, "Go ahead." So, I jump in the seat. I open the other side. The windows were blown out of the truck. It was all full of debris. Got the thing started, kicked out the emergency brake and everything, and started chugging and actually plowed through all the debris, cut around the corner, and took it down because they wanted to clear the street, because they wanted to bring dozers in [to] start pushing stuff out of the way and making room for emergency vehicles to come in. So, we ran this thing down the street, and when we got it down there, there was a refrigerated car. So, I said to my son, "I wonder what's in this thing." So, we opened it up, and [it had] water, juices, and everything else in it. So, I'm like, you know, everything was destroyed, and people were gagging. I said, "Let's start running this stuff up and bringing it in to the Burger King so they can start passing it out to the individuals." So, we actually—there was a Coca-Cola truck parked across the street from where I stopped this truck, and we took the hand cart off the back, and we were walking cases of soda up the street. When I get up

> the street and started doing this, I knew I couldn't move that much stuff by hand. This guy was like, "Great, can you keep this stuff coming?" I don't know who it was, but somebody said keep bringing stuff in.

After carting a few rounds of drinks, Wollman made a small discovery: he saw a police van, another vehicle that had taken its share of abuse, and asked a police officer whether he could use it to run supplies. He and his son emptied the refrigerated truck and then cruised up the street delivering more water and juice. Another officer approached them and asked whether they could head to 1 Police Plaza and pick up supplies, including masks and gloves desperately needed by those at the pile. Wollman recalled:

> I mean, this truck had no windshields, so we had masks on, goggles, and we're going, and we're hitting the siren, and we're flying up through the city, and cops are stopping traffic for us and letting us go. We got up there. We got a bunch of stuff. We brought it all back. I mean, they let us. 1 Police Plaza at that time was all barricaded in. They put road barriers all around it. So we showed up, and I guess they must have radioed ahead to tell them we were coming, because they let us right in, pulled right down into the garage. They loaded the van up with stuff, [and] we took off again.

They made trips all over the city. At a Consolidated Edison (Con Ed) facility that had some boots and gloves to send downtown, they piled the van with supplies, delivered them, and made another run. On the way, they ran into a contingent of National Guard troops trying to get to Ground Zero. They loaded them into the van, too. Making another trip to Con Ed, the van began to overheat: "We kept cases of Snapple and stuff in the van. We were putting Snapple into the radiator and juice just because we had nothing else, and, I mean, it was overheating big time. It wound up being because the radiator was so clogged with dust that it couldn't go."

On one of the trips to Con Ed, Wollman needed his eyes to be washed out. They were clogged with dust, too:

> We kept on running until like about 8:00 P.M. at night. We were making different runs to different places and getting medical supplies. At the last point, they had me go down somewhere by, by the ferries, and I guess there was ambulances down there with oxygen and stuff, and they loaded the van with different bottles of oxygen, and we made one more run.

During all this work, Wollman and his son relied on basic skills and knowledge that they recombined. There was a universe of jobs that needed to be done: they took some fundamental, mutable skills and found a niche for themselves in the free market of disaster. In that free market, practically anything could be useful; they needed no supervision and could work as freight haulers or envoys between different locations. They were inspired by their environment and could repurpose whatever resources were there. Importantly, officials did not obstruct them but rather welcomed them into their new and expanded circle. People found ways to help, afloat and ashore, by combining basic familiar knowledge and—piece by piece—building emergent organizations.

Emergent Systems

Imagine setting out to confront the effects of an ongoing calamity, an off-the-scale burst of surprise urban demolition, without knowing what you are going to do or how you are going to do it. Not only was there no advance planning; there was barely any immediate planning. Instead, "planning" consisted of a not-entirely-uncontested decision to do something and of a movement to collect the most mutable, durable, and useful tools and equipment available. Masks, water, towels, first-aid gear, and hand tools could be used for different scenarios that no one could anticipate even if their application was just a few minutes in the future. For a brief window of time in which everything was in flux, convergers had maximum freedom to take charge of their natural, social, and technical environments.

Lack of knowledge, lack of a plan, or lack of direction is hardly ever cited as a good set of circumstances in any situation—let alone a disaster. Instead, as we have seen, shared knowledge is important.

But more specifically, the timing and the content of that knowledge are key. In this case, not knowing their mission at the outset freed participants to do anything. They could carry passengers or transport supplies, or some of them, like Ken Peterson, could work ashore to build a traffic-flow pattern around Manhattan to get evacuees to the different embarkation points. Statements like "I didn't know what I was going to do" do not suggest an auspicious start for an attempt to move several hundred thousand people across the water, but they certainly demonstrate how much thinking and organizing can be accomplished right in the moment.

One person we talked with stressed the importance of getting "ahead of the game," suggesting that too much time spent thinking about a particular problem would lead to a series of reactions rather than actions: "I told the [person] that was with me, I said, [if I] get in the details, tap me on the shoulder and get me out of it. I didn't want to be solving every single little problem, because I thought I'd miss something major."

Certainly there were advantages to this approach. These convergers picked up the tools with which they were most familiar. They were working, at least to some extent, with people they knew well. And they were working in an environment that was simultaneously unfamiliar and that they knew intimately. Like an artist's reuse of a canvas for a new painting, where the palimpsest or shadow of the original is still somewhat visible, these convergers could rely on the familiar shadow of what had been there while adapting to the transformed environment.

But, of course, they did not have all the answers about what to do. No one could have had a full grasp of what was happening everywhere: the big picture. The people we spoke with improvised their tasks at hand with an assumption that others were understanding the rest of the picture. One person said:

> You know, there [were] people [who] asked us—how do you get to Newark airport? How do you get to the airport? "Well, either take that, or you go get a cab somewhere," it was like—well, is there a place for us to go and stay and get a hotel? I was like, "Well, yeah, you know." So you call up one of the little boats

> and say, "Hey, we gotta get these people outta here." But once we got them off the boats, I have not got a clue of what people did, how they did it, where they did it, who they did it with.

This person was busy with two tasks: getting people onto boats and keeping traffic flowing. He hoped that someone else would take the baton, that someone else had the answers about what to do next. He did not have to know everything, because elsewhere people were solving their parts of the problem.

Initially, many captains encouraged evacuees to get on board regardless of their destinations. One mariner said to the people he was transporting, "It doesn't matter where it's going. I'm getting you out of here. Get on the boat. . . . You got to get out of here and somehow, someway, you'll make your way to where you got to go."

And indeed, there were challenges once people arrived at a pier far from where they needed to go, particularly as the cell phone coverage was spotty that day. If someone had friends closer to Weehawken, arriving in Atlantic Highlands, some forty miles away by car, complicated the final leg of their journey home. P.J. and Bob Campbell humorously recalled one would-be passenger who took the prospects of being stranded across the Hudson River seriously:

> PJ: I tell it different from him. He tells it different from me. But it's the way I remember it. This little, like, ninety-year-old lady came out of the cloud—just gray-covered, with a cane. She's like, "Where you going?" I'm like, "Jersey." [She replied,] "Jersey? Fuck that! I ain't going to Jersey!" [She] turned around and walked straight back in.
>
> B: I was going to grab her.
>
> PJ: She would have whacked you.
>
> B: I don't care. I was going to put her on the boat. [And] then a New York City cop said, "Let her go." I was, like, "Really?"
>
> PJ: She'd rather die than go to New Jersey!
>
> B: Death before New Jersey!

Desire to help is not sufficient in disaster response. Beyond information and a goal, responders need some sort of "non-linear, dynamic system . . . to coordinate diverse resources, materials, and personnel across previously established organizational and jurisdictional boundaries" (Comfort 1999: 273). A free flow of information, with plentiful mechanisms for feedback among the various agents in a system, is necessary for maintaining a shared vision of shifting needs and the harmonized modulation of individual action.

As we have seen, plentiful information was available to the participants in the waterborne operation. VHF Channels 13 and 16—the marine "calling" and emergency frequencies, respectively—were open to anyone with a marine radio: private and commercial users, Coast Guard and other military units, police and fire craft all have such radios, which at least partially solves the problem of interoperability between preexisting systems. Everyone could communicate—indeed, excess chatter is often a problem in disasters and occasionally stymied communication on September 11, so that boat operators resorted to shouting or simply watching to interpret other vessels' movements. While these radios have a comparatively short range, it was enough in this circumstance. In addition to the radios, many of the participants were able to see through the smoke and dust to a broad expanse of the operational area—a rare situation in large disaster events.

The emergence of shore support operations in New Jersey is a good example of the sort of seemingly spontaneous responses to disaster that might better be described as a form of emergent organization. On the morning of September 11, Kim Newton arrived at her office at Sea Streak, a fast-ferry company based in Atlantic Highlands, New Jersey, after running some errands on the way in to work. One of the boat captains in Manhattan called her after the second strike, telling her to turn on the TV:

> You know I hadn't seen anything on the first plane yet, and that was the first visual I had of anything happening. At that point, we all turned around and looked at each other and just said, "What the hell's going on here?" One plane is one thing, but two is not an accident. At that point, it was probably a good five minutes, we just kind of all sat there a few minutes and thought, "What the hell do we do? What's gonna happen?

> Is there gonna be more attacks? Are we going to be able to go back in? What can we do?" So I would say within five minutes we had already decided to mobilize.

They decided to bring boats into Highlands, about two miles east of their usual docks at their office at Atlantic Highlands. She and one of the captains drove out there:

> At that point, we didn't know what other companies were going to do. We had not spoken to the Coast Guard. We had not spoken to the New York Harbor Unit. We just knew there was going to be a need for us. So we just kind of got everything mobilized and went from there. You know, Atlantic Highlands Police didn't know what was going on. Highlands Police, they didn't know what was going on. Nobody knew.

Geography—and, just as importantly, their knowledge of it—dictated how they would organize. Normally, their boats went into Atlantic Highlands, but this route involved maneuvering around a protective jetty that forms a little bay and that could add fifteen minutes to the docking and undocking process. The captains thought they needed a more straightforward approach, which they would have at Highlands. By about 9:45 A.M., Newton figured, the Sea Streak boats had returned to their New Jersey facilities. One captain was low on fuel. He had wanted to turn around midway and pick up the passengers he had just dropped off in Manhattan, but he did not have enough fuel to backtrack to Manhattan and then make the voyage down to Highlands. Another boat had turned around before disembarking its passengers when the second plane struck: "We had some people on the boat [who] didn't even know where they were going, and they didn't care at that point."

All the boats went to Sea Streak's facilities at Highlands. By now, local police and fire personnel were already there, having decided that one or two police officers and an EMT should be on the boats going to New York, to help with first-aid needs or security. While boats ran back and forth to New York, local emergency officials in New Jersey were busy setting up triage units in case other boats might be coming down from Manhattan. Midday, they had been asked by the Coast Guard and harbor units whether they could receive casualties. No one knew what might happen next. They did not know whether they would send separate boats for the walking well, injured, and bodies. In the end, all they received were the walking well, albeit quite shaken. As Newton described it:

> The first group of returning passengers were not hurt, just stunned and shaken up. The next boat that came back was the *Sea Streak New York*. Those people were covered [with] asbestos, dust, sheetrock, and debris, with burns. Some of their hair was singed, but they weren't physically hurt. . . . By the time that boat got back, the decontamination unit was set up by the county, and it was a sea of fire hoses, and they were just hosing everybody that walked off that boat. They got cleaned up from head to toe. In Atlantic Highlands, they had mobilized the Atlantic Highlands Office of Emergency Management team.

"And that's how we went," Newton said. "I mean, basically, went for the rest of the day, back and forth." They went back and forth to Pier 11, their usual pier, because their captains and crew were familiar with its configuration and because they knew it was the most likely place for their regular customers to appear. The crew changed in the afternoon, during which time they loaded up the boats with supplies for those evacuees they would bring to New Jersey or who were waiting for the next available boat. In the end, Sea Streak brought back about four thousand people, bringing the last load over to Highlands around midnight. Newton was not sure who arranged for the buses that started to arrive from several different private companies—probably local and county officials: "Next thing I knew, I was there and somebody said the buses were in the back of the lot. I hadn't known buses were ordered."

The next morning, September 12, Sea Streak's role changed from evacuating passengers to shuttling supplies to Manhattan. The company's operations manager heard a call on the radio for boots, socks, and underwear for the volunteers. He got in touch with the director of a children's foundation in Red Bank, New Jersey—about ten minutes away by car—who in turn got the word out on local New Jersey radio, asking people to bring supplies either to Red Bank or to the ferry terminal at Highlands. Now loaded up with material goods and firefighters, Sea Streak's captains radioed the Coast Guard, who at first denied them permission to come into New York Harbor. They explained what and whom they had on board, and they were directed to Liberty State Park. But when they arrived there, workers on shore saw how much material they had and sent them across the Hudson to Battery Park. Newton explained, "That whole area was a staging area. There were tents set up everywhere. You could see food tents set up. Red Cross. Everybody was operating out of that area." They unloaded their supplies but made only one run that day. Thursday morning, September 13, they did two more, bringing supplies dropped off by lines of cars driving into their parking lot. By this time, Sea Streak employees had also heard that welders and construction workers were needed at Ground Zero. Once again, they contacted local radio stations and put out the word that they could get these workers into the city. When a few—not many—workers showed up, one of the Sea Streak boats took them over to the Jacob K. Javits Convention Center at Pier 81. The next day, Friday, September 14, was rainy, windy, and blustery, and they made no runs to Manhattan despite all the goods still coming in. On Saturday, September 15, they did a morning trip.

The experiences at Sea Streak illustrate the idea of diffuse sensemaking and its connection to emergent systems. On their own, Newton and her colleagues gathered information about the growing crisis and made decisions about how they would react. Of course, some of that was concern for their immediate constituents: the commuters who ride their boats daily to Manhattan. But it was also generic helping behavior. "We just knew there was going to be a need for us," she said. Their activities became the nucleus for other work around the waterfront, as they got new information and reached out to the community. New activities emerged, such as collecting supplies from a queue of automobiles and then hauling these materials to Manhattan. All around the harbor, the process played out like this.

Gerard Rokosz, the general manager at the Lincoln Harbor Yacht Club in Weehawken, on the Jersey side of the Hudson River, described a similarly emergent process in managing the evacuees who found themselves stranded, some quite far from home. Lincoln Harbor is a full-service marina with 250 slips, providing docking space, fuel, water, sewage, and electric service to boats from around the world. It is a little more than three miles north of North Cove. Rokosz said that he watched the buildings burn for about ten minutes on the marina's security cameras before radioing the Coast Guard and letting them know that its docks were open and ready if needed. Maybe fifteen minutes after that, Mark Davidoff of Circle Line called, asking whether his company could bring boats over. Rokosz could not recall the time the first one arrived, but "it was way before the first collapse. The buildings collapsed while we were doing this."

Everything happened at once: "They brought a Circle Line boat loaded with people. We started marching people off." At the same time, Rokosz said, some people were trying to get through the gate in the other direction, toward the docks, to get a boat to New York:

> I had words with a couple of people [who] wanted to go back over. . . . [T]hat was the biggest thing, deciding if a guy, because a guy says [his] wife is pregnant, [he has] to get over there. I let him get on. Call the captain. He went back on one of the, uh, Spirit Cruises. I called him and said, "This guy's going on it. Take him back. You know, what happens, happens, you know, whatever he does, he does."

Meanwhile, people were disembarking from Circle Line, the tour boat *Horizon*, and other vessels. The marina was getting crowded. Local police were overwhelmed. But soon, Rokosz got a call from Academy Bus Lines: "They called me in because I had someone in the office managing the phones and they said, 'We know people are coming off the boats. We can't get down there.'" He called the police again: "You got to let the buses in. The buses . . . were picking up the people just wandering around here, you know. There's no way out." And so the police let the buses in, and the shoreside evacuation began: "They started letting buses go and the buses were pulling right up to the new building. You know, people were queuing up, very orderly, by the way. Getting in the buses, the buses were taking them to Giants Stadium or the Hoboken train station."

Marina staff shepherded people off the dock, through the gate, and along a walkway next to their office building, out to the street, "trying to be encouraging to [the] people," said Rokosz, "telling them . . . 'stay in line,' 'go to the street,' 'keep walking' . . . 'don't push.'" The first tower collapse broke the momentum: "Everything stopped. Everybody stopped, you know, the pier was full of people. The scene was dramatic as you can get. And, you know, they saw it. They can hear it, and everyone turned around and looked, and they were aghast. People were crying." And after this shocked pause, they continued to the buses, and boats kept arriving with evacuees: "It looked like . . . the retreat from Stalingrad. . . . [There were] women with babies, carriages, rabbis, guys who [were in] Brooks Brothers suits covered with dust." They wanted to know where the buses were taking them: "'New Jersey,' I told them. 'It's not that big a state. . . . You're not far from anything here. We'll get you home. Someone will pick you up. We're taking you to the stadium. Call somebody and get picked up.'"

At the risk of relying too much on metaphor, it is useful to consider the emergent response like one gigantic puzzle. Some people begin working on one part of the puzzle while others tackle the others, without a clear sense at the start of how the different sections fit together. Yet eventually, the connection or the picture becomes clearer. Ground Zero workers needed certain supplies. Supplies were coming into Highlands, and they could get them in. Eventually those operations connected. Boat captains knew people needed to get off the island but did not know what would happen to them from there. Others, including the bus companies, were able to recognize a solution. Here is how emergence continued to expand. It necessitated the linking of particular organizations, or networks of organizations, with others. What is particularly remarkable about such large-scale disaster response is that the participants working on the puzzle do not have a clear image to model their work after. They begin fitting pieces of the puzzle together without knowing what the big picture will eventually look like. Disaster is not a single entity; it is a collection of interpretations of how systems have changed and what needs to be done. For bus operators in New Jersey, it did not matter how people arrived to be evacuated—they could have flown over in an airborne evacuation. What mattered was what they saw in their corner of the puzzle: people who needed to be driven to Giants Stadium.

As Amico put it:

> The people that worked for the ferry company, everyone automatically did what they needed to do; no one had to tell anyone what to do as far as, say, it [might] be mechanics that usually repair the ferry boats [who] automatically hopped on boats to work as crew mates. The captains, as soon as they heard it, automatically came down, just as firemen and police automatically came down into the city. We are comfortable on the water where most of these office people are not, and we just had to do what we had to do. (South Street Seaport Museum Oral History)

Scattered across New York Harbor, the convergers read the environment and understood it through their identities and experiences—personal attributes that were shared along the waterfront and that formed the path for projecting ordinary life into extraordinary events. People defined for themselves where they could fit in, making a space in which to help that was just wide enough to be helpful but not so wide as to interfere with others. One person we spoke with said:

> What struck me down there was that this was being done very efficiently. And it seemed to me that things that needed to get done were getting done without a lot of hooey, and a lot of—not a lot of . . . worry. Just happened. Not a lot of organization from outside. Fascinating how that works. And it would be fascinating to know in which instances it doesn't. I mean, I assume if there's a sense of panic that that doesn't happen, but there was no sense of panic.

Indeed, widespread panic is rare. Much more frequent is adaptive helping behavior, especially in the early hours and days of an emergency. People help because they are right there; they have skills and supplies, even simple ones, that they can use before official responders can arrive, and they can bolster official efforts. They have a flat hierarchy, or no hierarchy, and they are free to flex and shift as needs demand. Sometimes they are the only responders on scene for a while—most people are rescued by friends and neighbors, for example (Aguirre, Wenger, and Vigo 1998; Alexander 2012). Sometimes these groups can fill in the interstices of more formal responses. Social ties and standards of behavior remain strong.

So far, we have emphasized what drew the different participants together: an identity as mariners or, in the case of the *Harvey*, as firefighters; a realization that a particular skill set was useful, as with Amico; and a recognition of a role within a community, even if people were itinerant in the community, as in the example of the Coast Guard. The convergers were motivated by community-based norms toward immediate goals, sharing knowledge and recognizing the limits of the knowledge and capabilities of others so they could fill in the gaps. They created a system that not only had never before existed but also had never before been imagined. When one hears of individual efforts, it would be easy to assume that someone is taking credit for the efforts of another. After all, how could so many people have initiated the same operation? But it is important to emphasize how many stories we heard of emerging joint logistics and supply operations that clearly involved coordinating the actions of many previously unconnected individuals in diverse roles. This is where the concept of diffuse sensemaking becomes particularly useful. Many people told us that the "big picture" of what was going on was not apparent to them until days or even weeks after the disaster. They were working in their own section of a complex and ever-changing environment.

Many of them had little interaction with operations taking place at other locations along the waterfront or other sides of Ground Zero, yet the issues they encountered were often the same. The supply operation in one area might have been precipitated by a need for water, whereas in other areas it might have begun out of a need for acetylene or fuel. Eventually, however, a waterfront operation to transport and distribute resources emerged. Sharing similar knowledge, similar physical and social cues, and similar identities, people made sense of their environments in similar ways and came to similar solutions to similar problems. We asked the mariners whether they would have done anything differently, and most said no. They would have liked better communication systems. They would have liked boats that were more easily amenable to boarding passengers. Two people we spoke with commented on their attire. One stated, "I just was in my business suit and didn't even think to . . . change out of my suit . . . but I would have changed into jeans." The other commented, "I was in a pair of shorts and a T-shirt. And that night . . . it got cold." Amazingly, given everything that was improvised that day, these were the problems that people came up with. Overall, it was a successful operation.

Once you make a system, you have to make it work. It is likely that the new system will depart from the norms and expectations of normal operations. This requires a willingness to not only make but also break rules, as we discuss in the next chapter.

4 /

Breaking Rules, Making Rules

The Paradox of Disaster

Lieutenant Mike Day of the U.S. Coast Guard was chief of Waterways Oversight for the Port of New York and New Jersey. He had spent eight months in a sort of exchange program with the Port Authority of New York and New Jersey, working in World Trade Center 1. The goal was to gain a better understanding of not only the maritime industry but also its relationship to, and perception of, the Coast Guard. The Coast Guard and the civilian maritime community usually have a love-hate relationship: the commercial mariners appreciate the Coast Guard's rescue function, but they resent its enforcement actions. Day's assignment was part of a series of recent efforts to build a better partnership, or at least a sense of shared purpose, between the Coast Guard and commercial operators in the harbor. The emphasis was on joint problem solving for maritime challenges in the harbor and reaching consensus on practices to avoid the need for regulatory solutions. Day was accustomed to performing a variety of water-based activities:

> A sea plane operator, for example, wanted to start a business cutting across the East River. They bought a tug and barge at scrap, so we were able to work with the sea plane operator and

> say, "You know what, if you choose to operate in this area, we're probably going to have to enact regulations that will put it to death. However, if you move your operation through another area, you can go ahead with your business venture, and you're not impeding the flow of traffic, and it's a win-win solution."

Thus Day was sensitized to understanding needs, opportunities, and perspectives of various maritime stakeholders and able to talk out solutions.

On September 11, Day had just entered a daily 8:30 A.M. briefing at Coast Guard headquarters on Staten Island. He provided oversight for New York Harbor. Shortly after the meeting began, a watchstander came in and reported that a small plane had struck the World Trade Center. Despite initially assuming that it was the crash of a sight-seeing plane, the Coast Guard sent a forty-one-foot patrol boat to check on the situation. Monitoring the vessel traffic control cameras overlooking the harbor and watching CNN, Day initially thought it was a nonmaritime event, or at least one that would not escalate to a major water-based response. "There's so much going on in New York. I mean, we're so used to barges exploding or something," he said. Of course, when the second plane hit the second tower, Day and the rest of the Coast Guard knew the situation was more serious.

About that time, Sandy Hook harbor pilot Andrew McGovern arrived at the Coast Guard base. McGovern chaired a harbor safety committee on which Day served as well. He was also on his way to attend a meeting of the committee in Manhattan at 10:00 A.M. He was driving in from his home on Long Island:

> I heard a traffic report. The traffic 'copter said, "Oh, there's a fire at the World Trade Center!" and the helicopter was literally—you know—he was going around the Trade Center and saying, "They say a plane hit it, but it must have been a small plane, 'cause it didn't come out the other side." I remember those words: "It didn't come out the other side."

In fact, McGovern could see the Trade Center on fire from where he was driving. "Son of a bitch, now it's gonna screw up traffic," he thought. He could not have anticipated the redirection this day and the coming weeks would take.

It is difficult to reconstruct the exact order of what unfolded next. The accounts we examined varied, but at some point McGovern connected with Day. Decisions were made all at once, both by group and individual deliberations, blurring the boundary between individual and collective contributions. At some point, people began discussing when to issue a call for "all available boats." Day was chosen to go to the area and was told to marshal a few people to take along. He pulled in Chief Petty Officer Jamie Wilson, six petty officers, and a civilian maritime inspector, Ken Concepcion. Wilson was a Chief Boatswain's Mate, or a BMC. In the U.S. Navy and the Coast Guard, boatswains are the expert mariners, the boat operators, the ones who know the fine points of seamanship, or the art and science of using nautical equipment. Wilson had been stationed in the Coast Guard's small boat station in New York Harbor prior to coming to work for Day; he knew the harbor from an operations perspective. As a maritime inspector, Concepcion knew all the capacities of the vessels in the area, so he was also a good choice to recruit.

In addition to a crew, Day needed documents. He pulled together harbor charts as well as a copy of a plan for potential staging areas for emergency medical care originally developed for an Operation Sail (OpSail) event the previous year. During the OpSail, the city hosted a large parade with tall ships, warships, and yachts from all around the world. The OpSail plan was the closest blueprint that the Coast Guard had for dealing with a large maritime emergency in New York Harbor, and it was for medical and logistic staging procedures. Remember that no plan existed for a boat-operated island evacuation. Not knowing exactly what might happen, Day thought that the OpSail plan would be a good document to have on hand. Day's colleagues, meanwhile, were reviewing plans that had been drafted to provide assistance for a capsized Staten Island ferry in winter temperatures. Those documents provided a rescue plan for a worst-case scenario: how could up to six thousand people be rescued from freezing water and winter winds before they succumbed to hypothermia? Neither plan provided detailed guidance for what became the waterborne evacuation of Manhattan, but each furnished some of the raw material for fashioning an emergency response from the emerging civilian, law enforcement, fire, and maritime personnel and equipment that would pour into Manhattan over the next days.

As communications started to give out, Day and his colleagues began to realize they would need to take a more-flexible-than-usual stance on harbor rules. One of the most pressing problems was coordinating harbor traffic, as ferries, water taxis, tugboats, and private operators started to undertake their own rescue missions. The Coast Guard started by sectoring off channels on the ship-to-ship VHF radios for different groups, but McGovern realized that they would need more durable communications and a recognizable command post. He contacted Jack Ackerman, a Sandy Hook pilot who also served as the marine superintendent for the pilots' headquarters, and asked Ackerman to bring over the two-hundred-foot pilot boat *New York* to Manhattan. "Pilot boats are basically floating command-and-control structures," McGovern explained. "So they have huge bridges for their size, 360-[degree] views, and lots of radios, and lots of radars, and lots of everything. So it's really a perfect platform, so we volunteered that."

In a fairly major departure from Coast Guard operating procedure, Day decided to use the *New York* as a base. He made the decision to hoist the Coast Guard's flag. "I wouldn't say [we] deputized it, but I took it over," he clarified. It was not really a matter of fully taking over, but the pilot boat with the Coast Guard flag symbolized the cooperation between the two entities that would follow throughout the rest of the day. Certainly the *New York* was well equipped as a command facility, but Day also recognized the immense amount of knowledge that the Sandy Hook pilots brought to the table. Day anticipated that the harbor pilots would offer a better view of the situation than other Coast Guard officers his own age, because of their experience with a vast range of vessels in the harbor and because of their relationship with the maritime community. In addition, he himself had a strong working relationship with them. At the same time, McGovern recalled asking for some Coast Guard personnel to accompany the pilots: "I said, 'We'll need some Coasties.' We needed the uniforms 'cause . . . people will respond to a uniform more than they'll respond to nonuniformed personnel." Each organization had something to offer, combining strong local knowledge and personal networks, institutional authority, credibility, and equipment.

With communication systems periodically down, putting up a flag was the easiest way to communicate the *New York*'s new coordination mission to other vessels. As much as the burning towers or the residual plume of debris overhead signaled an unprecedented environment where unconventional steps would prove necessary, other symbols and visual cues repeatedly served as signals for new activities, directives, and operating procedures. Such signs served a practical purpose, but they also pointed to the urgency of bending or breaking rules and making new ones to deal with an unprecedented situation. Performing functions in new ways was now part of the evolving environment.

Rules for Maritime Risk

The maritime life is a paradoxical one, a blend of high technology and folkloric best practices, hierarchies and regulations, extensive planning and training, and improvisation and making do. Running a ship typically relies on rules, procedures, sequences of tasks, and checklists. These are drilled into prospective mariners through long periods of training, either at one of the various maritime academies, in union-sponsored education programs, in military service, or on the job. Dropping an anchor, lowering a lifeboat, testing engines and steering before getting underway, and making a ship ready for sea are all tasks accomplished in an accepted order that a seafarer is expected to know and perform with an automatic precision and blended rhythm of thought and movement.

Aphorisms and age-old adages prompt the management of different functions, so that sailors learn, for instance, that they are "never far from a good DR"—that is, their dead (deduced) reckoning position, an estimate of their future position based on their current course and speed. This common saying serves as a reminder to keep careful track of a ship's course and speed (and to account for wind and current) and as a check on fallible navigation aids, which are never to be trusted. Mariners learn that "the prudent mariner will not rely solely on any single aid to navigation, particularly on floating aids." Indeed, this important reminder is printed on every nautical chart. Mariners learn that "red-right-returning" means to keep the red buoy on the starboard side approaching a harbor. The phrase "hot to cold, ventilate bold; cold to hot, ventilate not" reminds operators how to keep cargo from getting damaged through sweating when sailing between climatic zones. Customs, norms, and mutual expectations among mariners, propelled by literally hundreds of years of seafaring tradition, underlie every order, act, or transaction in the maritime world.

One of those traditions is an appreciation of hierarchy. The first pages of the venerable *American Merchant Seaman's Manual* (1980), the preeminent handbook of seamanship, contains an extensive description of shipboard organization to initiate the newcomer to the industry. Even the efforts of modern industrial engineers and workplace psychologists to remake the shipboard hierarchy in a more "participatory" model have been only partially successful across the globe. For example, cadets at Massachusetts Maritime Academy, during their first-year indoctrination, are taught to recite a paraphrase of Alfred, Lord Tennyson's (1854) famous lines, "Sir/Ma'am. Mine is not to question why. Mine is but to do or die. Mine is not to question how. Mine is but to do it now, Sir/Ma'am." At the maritime academies, cadets, or midshipmen, as they are sometimes called, spend their four years in a naval-style regimented structure that highlights obedience, accountability, and authority, rationing out progressively more responsibility as the students proceed through the program.

Just as there is an individual hierarchy, there is an institutional one. The U.S. Coast Guard is the primary regulator of the merchant marine in the United States. It develops and enforces regulations for the training and licensing of sailors, for shipboard work and living conditions, and for standards for ship construction, machinery, and equipment. The Coast Guard is the principal enforcer and looms large in a sailor's career, since advancing one's license to the next level requires passing a Coast Guard exam. The Coast Guard is the main purveyor of some thousands of pages of rules contained in the Code of Federal Regulations, covering every aspect of maritime life, from the subjects to be tested on a master mariner's examination, to the required equipment aboard a lifeboat, to the minimum height of a handrail above the deck (39.5 inches).

But in spite of the emphasis on rules, accountability, and procedures, the extent to which rules can be grafted onto the unpredictable maritime setting is often limited by either economic or practical circumstances. The Coast Guard regulates for safety, but it cannot regulate *too* much, or else it will begin to impinge on the competitiveness of U.S. ships—the same regulatory dilemma encountered in all hazardous industries. The International Regulations for Preventing Collisions at Sea, known informally as the "Rules of the Road," guide the interactions between vessels, choreographing the collision-avoidance dance of ships at sea or in a crowded harbor. Navigation officers memorize pages of rules for meeting head on, crossing at an angle, overtaking, and the complications of sailboats tacking, narrow channels, and fog. The object of the dance is simple—do not collide—but the rules are really only a starting point for the adaptive process of a lot of ships negotiating their way through a busy waterway. Often the rules do not cover a given situation, but there is even a "rule" for that—traditionally known as the General Prudential Rule and now simply codified as Rule 2(b), it requires that mariners deviate from the rules if necessary to avoid danger. Rule 2(a), the Rule of Good Seamanship, cautions sailors to make sensible decisions based on "the ordinary practice of seamen"—that is, on the art of seafaring.

Knowing when to break a rule—when to invoke Rule 2(b)—involves an assessment of risk. Bound to obey rules on the one hand, and, contrarily, bound to see past rules, mariners make their own assessments of growing dangers and peculiar conditions that may be developing. They are frequently left to their own judgment in moments of danger where there is often little time for sustained or careful deliberation. Noël Mostert (1974: 155), in his classic *Supership*, quotes a tanker master:

> I've never forgotten a board of inquiry on which I once sat. A British India Line master was up before us. His ship had been in a collision. They asked him why he'd given the orders he did when he saw the danger he was in, and he simply said, "I had thirty seconds in which to make up my mind and I did at that moment what I thought was best." There was no answer to that, and they let him go.

Sometimes an assessment of risk requires that a mariner break a rule, but in other cases, following the rules seems to expose a mariner to additional risk. A notorious example is the Texas Chicken maneuver, used most famously in the Houston Ship Channel but in other waterways as well. In particularly narrow channels, two ships are not able to pass simply alongside each other—hydrodynamic pressure effects would force each ship into the opposite bank. To complete the passage, the ships steer directly toward each other and turn aside at an agreed-on instant. The bow washes force the ships apart. As the ships pass, each ship turns toward the other's stern, where suction pulls the bow in, away from the side of the channel, and the ships proceed on their courses. But when asked about it, a pilot in Houston downplayed the risk, citing the training and experience of the pilots; he resisted the appellation of "Texas Chicken," which, he said, journalists had applied to the maneuver (Kendra 2000). Such a maneuver plainly pushes a variety of margins, but the Port of Houston relies on it to maintain efficient traffic flow, explicitly exchanging risk for economy (National Research Council 1994). Pilots, thus, are obliged to take some risk, and to bear personal and professional risks, to maintain the functioning of that port. And while technically doing any job carries risk, having a channel that cannot accommodate two ships side-by-side suggests a social preference that the risk be borne by the individual. Mariners are used to absorbing the risks of other people's preferences.

Kendra (2007: 31–32) observed:

> Mariners occupy a problematic conceptual space within the shipping industry. One of their many duties is the identification and reduction of various sorts of operational hazards in the complex marine environment. Through the application of skills learned in formal and informal training and bolstered by experience, merchant mariners confront hazards posed by concentration of shipping traffic, congested ports, narrow channels and shallow water, and extremes of weather; all against the backdrop of high tempo operations and tight schedules.

Seafarers' lives of contradiction may explain why the mariners who were involved in the boat evacuation on 9/11 slipped so easily into this situation with such ambiguous guidelines and roles. It is their professional imperative to take and manage risk. They are compelled to trust certain people and technologies and to distrust them if they fail. How these conditions fed precisely into each mariner's thoughts on 9/11 is impossible to say, for mariners do not talk in terms of conflict or contradictions. But the idea is apparent in their tone; they wanted to do what is right but not get into trouble for it. Their jobs demand alertness for the unexpected that stems from contradictory principles, such as the need to doubt the information provided by global positioning systems (GPS); their jobs demand a continuous navigation of ambiguity. It seems reasonable that comfort with instability made it possible for these sailors to head toward Manhattan without a clear sense of what they were going to do, equipped with their usual tools and a willingness to act spontaneously on what they saw, even if it meant breaking the rules. They were living Rule 2(b).

A Changed Rule-Scape

The regular rules of the harbor fell by the wayside in fairly short order—but, interestingly enough, to only a certain extent. Day's decision to fly the Coast Guard's colors over a harbor pilot's boat may not have been standard operating procedure, but the move served to tell boat operators where they might look for guidance for the *new* rules. This pattern would repeat itself throughout the day.

By the time that Day and his ad hoc crew of pilots and Coasties started heading out toward the tip of the Battery, he had already started getting reports of boats picking up evacuees. He soon lost VHF radio communications, but back in Staten Island came reports that many people were gathering along the lower tip of Manhattan after the first tower collapsed. As Day approached the Battery, he recalled, "There were just boats everywhere. Tugboats, ferryboats . . . all different dinner cruises of every kind." As impressive as the response was, Day's eye was for safety, because that was part of his job: "Everyone wanted to help, but it kind of doesn't make sense since they were in the way. So, I used this forty-one [foot patrol boat that had arrived] to kick people out that were too small. I mean, they practically were ma and pa kind of [boats]."

This sentiment was reinforced by several people we spoke with. Had the disaster happened on a Saturday morning when they could have expected hundreds of small recreational boats, the challenges of coordinating the evacuation might have been even greater. One person said, "It looked to me like an accident waiting to happen. They were just there maneuvering all over the place. . . . Big boats and small boats don't mix, in my opinion." Another recalled, "It was kind of haphazard. . . . [W]e had tugboats going to the ferry docks and ferries trying to go to the Battery Wall." Looking at boat size and capacity was the first step in sorting out the boats that were maneuvering all over the harbor. One of Day's first acts was to ask all boats wanting to assist to come up to a staging area at Governor's Island. The pilots knew the vessels just by looking at them, and some of the smaller boats were redirected elsewhere. "I didn't know I was going there to do an evacuation. . . . I was sent there initially to observe," Day confessed.

Since it was not necessarily an advantage to have a working radio, with so many people talking at once, Day started putting vessels on specific routes. Until they heard otherwise from him, he directed the vessels to just keep going back and forth along those routes. Sometimes he identified a specific spot; other times he directed them toward a general area and instructed them to unload wherever they would land most quickly. Sometimes, the vessel size determined its drop-off point. The huge Staten Island ferries, for example, could move thousands of people at a time and had specially constructed facilities. Since they could not dock anywhere else, that was where they went. In other cases, the goal was to find a location that would keep the travel time down to a minimum. It was a key decision, according to Day, to resist communicating with every vessel each time it crossed the harbor. Instead, he essentially put the mariners on autopilot until they heard otherwise from him, because, as he noted, "I couldn't keep up."

Pier 11, on the eastern side of Manhattan, was used for some ferries, while the Battery Wall was used for the tugs. North Cove and South Cove were also used for embarking passengers. New York

Waterways used its own terminals along the Hudson River. By the time Day and McGovern had arrived, the largest congregation of people was at the Staten Island Ferry Terminal. Day knew the ferry crews needed help, and he looked to the OpSail plan in hopes of identifying alternative staging areas.

A primary goal was to get any injured to an ambulance staging area. Day said he heard reports of ambulances converging to different spots, and this stuck in his mind. He passed the word along regarding where these injured should be taken, but not many needed that kind of help. A few people had secondary injuries—from falls or accidents while fleeing—but not many had been injured at the actual World Trade Center site. It was only later that all of us, there or not, would grasp the fairly sharp line between walking (or limping) away and not making it out alive at all. As Day and his colleagues moved to different locations, they quickly saw which emergent efforts—including rule making—were working and which were not.

When the Coast Guard issued its call for "all available boats," it was nominally in charge. Wilson and Concepcion were dropped off to help organize efforts at different points along the shore. The boat operators tended to quickly assume an evacuation role, regardless of whether they had come on their own or in response to the Coast Guard's call. Ken Peterson, who was spearheading Reinauer's efforts, was in touch with the *New York* over the radio; since several of the Sandy Hook pilots knew him from various harbor operations and safety meetings, he had no problem getting authorized to continue. One harbor pilot described the effort at the Battery and the coordination among the tugs as "a decentralization kind of thing. If [Ken Peterson] needed anything, he called us." Indeed, some mariners on the tugs had no sense that the Coast Guard was even involved in the effort along the Battery. As one mariner put it, "Maybe people assumed there would be somebody on the radio or maybe even a city official or a Coast Guard rep, since the Coast Guard had put out the request for help. They would be waiting in Manhattan or even communicating from Staten Island at the command center. . . . There was none of that." Another tug operator noted, "You assume that somebody is summoning you for a reason and that they have a plan. . . . I can certainly appreciate what was going on. But [the Coast Guard was] more than accommodating [by] letting [others on the waterfront] coordinate it."

In reality, the Coast Guard had dispersed its authority to other knowledgeable parties. One person described Day, in particular, in this way: "I mean, and he didn't strong-arm anybody [or say,] 'I'm the Coast Guard officer. I'm in charge.' He just went with the flow. 'If you need me, I'm here. . . . ' It wasn't like he was trying to play Big Mr. Coast Guard. He was a member of the team." The Coast Guard and the pilots had a hand in directing efforts, but they also let the process unfold and knew that activities that were working were already underway.

Playing with the Rules

One mariner we interviewed said that at first, he just responded to the needs at hand: people had to be moved, supplies needed to be transported. He said, "I think I was numb. I didn't know how to react, but I just reacted." He made a lot of fast judgments about how many people the boats could safely carry, not counting the number of passengers but instead relying on a seaman's eye to estimate the seaworthiness of the boat and the safety of the operation. Another mariner said, "It's just activity where you're running on adrenaline, you know. You're not rationalizing everything, but you're moving." As time went by, though, and the first shock and surprise diminished, the evacuation gathered its own momentum and its own emerging routine. This loose organization, informal and improvised as it was, was gradually learning something about its environment and its capabilities. Everyone started to settle down. Reflecting back on it, one boat captain said, "It got to a point where it was like you started to bounce back to reality. Let's slow down. Let's not get stupid here. Let's not end up with a bigger situation that doesn't have to be."

Some of the trepidation came from the overloading of boats, which took place more often in the initial hours of the evacuation than later in the day. Most of the mariners we spoke with conceded that boarding more people than the boats were regulated to carry did occur, sometimes to too great of an extent, as the captain quoted above alluded to. Yet many of the people we interviewed also detailed the care, knowledge, and reasoning that went into this deviation from the number of passengers they were authorized to transport, at least after their initial trips.

It was hard to control the number of people getting onto the boats, and the crews were not much in the mood to restrict them as vigorously as they would have if these were passengers embarking for a normal tour around the harbor. People crowded the piers farther north and the paths and walkways along Battery Park. As soon as they were permitted to do so, they rushed onto the boats:

> Well, we didn't want to overload, but we didn't want to just take the exact amount we needed. So we had clickers and the guy would stand at the door as they were coming on and then he was clicking them on, you know what I mean? So I'd say, "How many you got?" He'd say, "200." All right keep going, keep going. You saw so many people waiting to get on. He didn't want to say, "Okay, only put 220." So we put on 250, we put on 300 and said, "Okay, that's enough." You didn't want to put, like, 600 on it, or you'll have people turning the boat over. . . . We took as many as we could possibly take and [still consider] that it was safe, and then we went.

Crews lost count of the number of passengers. As one mariner put it, "It only took us like ten or fifteen minutes of getting everybody on. . . . We lost count of them. They were just coming in so fast." But in most other cases, the boats were deliberately, although judiciously, overloaded. Passengers were counted, and captains determined what they assessed as appropriate cut-off points. As the *Ventura*'s Patrick Harris describes in the South Street Seaport Oral History, "[The captain] was sticking his head out the window of the bridge and he was saying, 'Cut it [the passenger loading] off at about 300.' He knew that he was going to break the law, but nobody was going to [criticize him]. So about 300 people were put on board." This was a risk they assumed, balanced with the need to get people off the island.

One mariner we talked with recalled taking on one hundred passengers when his boat was allowed to carry only eighty. The crew told those on board, "Sit down and do not move. We don't want the boat to capsize." Initially, that worked. Later, however, they briefed people before they even got on board rather than waiting until they were seated. Another mariner made a similar statement: "When we first started out, they were putting as many people [on the boats] as possible, so more than likely they went over the limit." But as the operations continued, the mariners increasingly brought back into their repertoire more of the basic guidelines.

Some of the larger tugs could carry a number of passengers, even if they were not precisely built for it. The *Janice Ann Reinauer* capped the capacity at about fifteen to twenty people. The crew cleared away as much of the towing equipment and wires and other hardware as they could to eliminate tripping hazards or obstructions that might potentially hurt people. For example, Kevin Tone told passengers where to sit so they would be safe on a vessel not designed for carrying more than a small crew. He gave a safety briefing before the tug got underway and showed passengers how to put on and take off flotation devices. He told them what the vessel was configured for, how long it would take to get to the other side, and where to hold on. He asked whether they had questions. Clarity and safety were paramount, even—perhaps especially—when they were technically breaking the rules. They were lucky, of course, in that the weather was perfect, with moderate fall temperatures and not a cloud in the sky. "Thank God for that," Tone remarked. "So the people weren't going to be exposed to the elements out there. Shivering or shaking or being rained on. Or it wasn't overly hot, so we [didn't have] to worry [about] heat . . . exhaustion or anything like that."

Boats sometimes used piers they were unaccustomed to or that were not designed for passengers and had to jury-rig gangways because of the different heights. Tugboats, for example, have no obvious entry points. Crews on many tugs improvised steps and gangways from what they had on hand, but they had to exercise caution with passengers using crutches and walkers. In fact, many of the boats involved lacked gangways; ladders were rigged, or people jumped into the boats, being half-caught by the crews or half-tumbling onto the deck. Here, vigilance substituted for the proper equipment. Remarkably, no serious accidents or injuries occurred during the evacuation, beyond one person breaking a leg. We heard conflicting reports of heart attacks among the mariners—perhaps one, perhaps three. We have not been able to confirm how many, but the limited evidence suggests that these were stress-related rather than a function of specific decisions or situations.

The possibility that volunteers will do damage or cause more trouble than they are worth is a main concern among emergency officials everywhere, including on 9/11. The concern is not totally misplaced; if people try to tackle jobs they have no training for, they can make a bad situation even worse. In this case, however, some eager operators recruited passersby without maritime experience or credentials. And, of course, technically, the mariners were *themselves* volunteers. One mariner highlighted the importance of knowing your overall strengths in the improvised effort: "We had to make do, because this is not something you're normally trained for. Remember, we're just private citizens. I was able to draw on some of my military training and my Coast Guard experiences. But some of these people, in the maritime industry, these guys are all like, you know, dishwashers [who] hire on as deckhands." Certainly that wasn't true of all deckhands, and we interviewed several with a considerable amount of training and experience. At the same time, this mariner's point was that "you have to know what your limitations are with the resources you have, including the people," and make do with what is there. Sometimes, this process violates the rules. In times of crises, however, people may see breaking the rules as the best option, certainly better than doing nothing, and they incorporate volunteers into their activities—albeit with caution.

There was, nonetheless, one significant rule that was not broken: mariners have an obligation to rescue. Many of the sailors we talked with—indeed, most of them in one way or another—described their roles in terms of inevitability. They saw no real choice on *whether* they would try to help; they simply needed an opportunity to fit in and some latitude to take individual action. That sense of inevitability—that only one course of action would really be appropriate—is part of a larger cultural bond in the maritime community that extends beyond this particular time and place. Many of the mariners we talked with emphasized their professional duty to assist someone in distress, citing the long-standing traditional and legal obligation well known even to nonmariners. Clearly, this was a powerful motivator, because professional honor was at stake. Yet upon closer scrutiny, the obligation to rescue does not provide the full underpinning for what the maritime community did that day. According to U.S. law:

> A master or individual in charge of a vessel shall render assistance to any individual found at sea in danger of being lost, so far as the master or individual in charge can do so without serious danger to the master's or individual's vessel or individuals on board. (46 U.S. Code Section 2304)

International treaties assign the same responsibility. Whether that same obligation exists to assist someone in danger on *land*, however, is more ambiguous. Some boats, such as fireboats, fight fires ablaze in shoreside warehouses, but it would be difficult to imagine a lot of boats rushing up for traffic accidents or other shoreside emergencies, even though those situations also place people in danger. On 9/11, however, the boat operators, and, indeed, other workers, stretched the boundaries of their obligations beyond the water to encompass some part of the shoreline. Those who work in New York Harbor are intimately familiar with the waterfront and see it, at least in part, as an extension of their usual environs. In contrast, deep-sea mariners, who are often strangers to the places they visit, look at the shoreline with suspicion. It represents the danger of shallow water or submerged hazards. The same call to respond might not have resonated for those mariners. But for inland sailors, especially those who live in New York or New Jersey, it did.

Here, the mariners in the harbor were applying a rule—the rule to rescue—that, through its very technical interpretation, did not actually apply in these circumstances. They came back to this explanation time and time again, but its appropriateness may have also been rooted in that idea of protecting a place to which they felt a homelike affinity.

There were other examples of making and breaking rules. One example demonstrates that even when rules are made, they do not emerge in a vacuum. Rather, they are a result of drawing on existing knowledge or experience, interpretations of the present circumstances, and the application of knowledge to those interpretations. Popular culture and historical references from books, movies, and films proved to be important in shaping the mariners' understanding of the situation they were in and what was required. Most of us are familiar with the "women and children first" dictum in an abandon-ship situation. The reference stems from the sinking of the *HMS Birkenhead* and, of course, is more familiarly immortalized in the history of the *Titanic* and its many representations in books and films. It is a compelling cultural norm. The Sea Scouts of the Boy Scouts of America have it in their motto, although it is not accepted as a recommended practice in U.S. maritime training.

Aboard one boat on 9/11, deckhands shouted to evacuees on the dock, "Women and children first!" We were fortunate to be able to locate and talk with these mariners. They recalled examples of how they organized embarkation, such as letting an older couple on board together and a young couple with a baby. But men who arrived on their own were asked to wait. They even queued women along the wall for protection, should another building collapse, and queued men along the railings. They explained the edict in several ways, at first stating, "In a maritime situation . . . the boat's going down; it's women and children first." They then pointed to their military training. Finally, one admitted, "I think we've seen too many movies."

Yet we know that this "rule" of women and children first is, in fact, not part of any official or codified modern maritime evacuation protocol. On cruise ships, for example, members of the same cabin are typically assigned to the same lifeboat; the only exception is that if children are involved in activities away from parents, staff escort them to their assigned station (Holland America Line 2015). In other instances, "if a child is not picked up, a Child Care Team member will accompany the child into the lifeboat to ensure they are accompanied at all times" (Ogintz 2012, citing Anne Marie Mathews). Although an individual cruise ship company could, of course, generate its own ad hoc procedures, typically no mention is made of women and children having precedence in evacuating ships or planes, as most of us know from pre-takeoff instructions. When and if such activity occurs, it is as an "emergent norm," a new behavior influenced by the environment and guided by historical references and gender relations, not by official maritime practices.

Evacuation research tells us that people tend to evacuate in social units (Aguirre, Wenger, and Vigo 1998) and will delay their evacuations until the units are reassembled. Those social units may comprise family members, but they may also comprise friends and coworkers. This emergent boarding procedure does raise important questions about how large-scale evacuations take place. Few people could find fault with the mariners recognizing the susceptibility of the very young, the very old, the injured, or those with mobility limitations owing to the health hazards of the dust permeating Lower Manhattan. Few people would not recognize, if not necessarily appreciate, the chivalrous instinct demonstrated by allowing women priority in boarding. At the same time, is there an advantage to keeping back a male coworker while his three female colleagues board a boat? Is there an advantage to separating two strangers who have found comfort in each other while evacuating the towers? It is one thing to ensure an orderly embarkation process for boats precariously hugging the seawall; keeping single men behind is quite another. As much as physical and social cues, identity, and past experience influence how we improvise, so too, we would argue, do other sources of knowledge and the frames through which they are interpreted. Folk and popular culture representations of boat rescues, interpreted through a mariner's lens, arguably influenced the "women and children first" rescue protocol that morning.

Other norms emerged that would have been unthinkable on September 10. Several people we spoke with mentioned disregarding company organizational lines of authority. Vessels that normally took orders from their own company dispatchers, for instance, began following directions from people on the waterfront who were coordinating the evacuation—and the companies seemed fine with that. Another obvious instance of rule- breaking was in scavenging necessary supplies or equipment, as Jeff Wollman did when he offered to hot-wire a vehicle for a policeman and distributed drinks straight off the back of a delivery truck.

Wollman's actions might be interpreted as those of an individual operating on his own initiative, but scavenging became part of the semiofficial supply chain. We are not talking here about widespread looting—extremely rare in most disasters—by which people are trying to enrich themselves. Here, mariners and waterfront workers commandeered property for what they defined as an urgent need. One harbor official recalled coming across a firefighter near Ground Zero who said they desperately needed drinking water in that area. A radio call went out, and soon hundreds of gallons of water were being transported by boat to the shoreline. But how to get it to the firefighters near the World Trade Center, a few blocks away? Fire hoses, pumping water from the river, blocked streets. Even if they could commandeer a truck, driving it up that way would cut off the water supply for the fire suppression. As Wollman did at first, someone else tried using a dolly from the concierge area of a hotel, but this method proved inefficient, to put it mildly. One of the harbor pilots considered how supplies could be shuttled more easily:

> [We saw] that there were two golf carts in the dock, and they were tied up right in front of the police boat. So we went over and we talked to the captain of the police, and we asked him to please turn around because we were about to liberate two golf carts. So we took those golf carts and put them to good use. And, of course, I live in Jersey, [my friend] lives in Jersey, and not being city kids, we didn't know how to hot-start these. So we got one of [the] engineers who could get it going for us, and then we started bringing supplies well up into the rescue site for these people. And it was great, because the golf carts let you go from the north side of the World Trade Center area down through the south side, the west side, and somewhat over to the east side. So we [were] able to be pretty far-ranging with where we could bring supplies [from the water operation] for people.

The use of golf carts near Ground Zero took on a life of its own. Eventually, golf carts came in from Monterey Golf Course on Staten Island, Chelsea Piers, and other locations. These golf carts inspired a fresh innovation: a week later, McGovern was able to get John Deere to send down similar four-wheel-drive vehicles from Canada. Over the coming weeks, those Gator utility vehicles became a well-known means of transportation around Ground Zero.

The people we talked with were not especially prone to criminal mischief, such as "liberating" and hot-wiring golf carts, and there is a certain good-naturedness to asking the police officer to turn around while they did it. They redefined what was acceptable. In some situations, volunteers became so well-organized that officials saw little choice but to let them proceed despite the fact that the operation resisted the authority of formal organizations.

Allan Barton (1969: 38) has written that communities dealing with "collective stress" situations become, at least briefly, *altruistic communities*. But when altruism is obstructed by officials, the altruistic community becomes a resistance community, where the helpers resist authority in order to do their work.

One such example involves volunteers working at the Jacob K. Javits Convention Center. This tremendous spontaneous volunteer effort sprang up in the first days after the attacks.

The Mayor's Office of Emergency Management had been driven from its emergency operations center, which burned throughout September 11 and collapsed late in the afternoon. For a couple of days, its focus, and the focus of other city agencies and principal disaster organizations, such as the American Red Cross, was on establishing a functioning and properly equipped facility to handle communication and information processing. During this time, multitudes of potential volunteers were left more or less on their own. When left without information or direction, people do not sit passively until told what to do—they will begin to take action if they have the resources. And after the attacks, volunteers and supplies were pouring into the city in superabundance. The Javits Center was designated as a marshaling point, but once they got there, the volunteers did not just hang around waiting for orders. At one point, as one official described, two separate response supply operations were underway:

> There was the official sanctioned operation, and then there was the unofficial unsanctioned operation. And both were taking stuff to the same place. . . . That was their way of contributing to the effort. They really weren't getting in our way.

> They weren't doing anything negative. They weren't breaking any laws that we knew of. They weren't creating any unrest. But the only thing that you could say is in the end they were using supplies that we wanted to use.

Volunteers at the Javits Center identified what they believed were unmet needs and took it upon themselves to get supplies to Ground Zero through unofficial means. In the end, as the convention center wanted to return to normal operations and as the official response tried to gain greater control and oversight, incremental steps were taken to disband the unsanctioned efforts. Subversion of emerging rules, such as who was in charge, contributed to how long the unofficial supply operation functioned on its own.

The same thing happened along the waterfront. Where the supplies were coming from, few people knew. "There was tractor trailers full of water backing up to the dock . . . and a crew showed up with boxed lunches, you know. It was pretty amazing," one person we talked with recalled. People organized themselves; they sorted and categorized donated supplies. There was enough work to go around, and the waterfront workers stepped in to do it according to their best judgment. If they needed something that was available, seemingly abandoned, to help the response, they took it. As one person we spoke with explained, "Many of these milestones were actually self-conceived—I mean, there is no template. There is no recipe. There is no plan to choose from . . . [no] menu to pick from, to decide what to do next. . . . You probably know what they need and you figure out a plan to do it." They made up those plans, and the rules to go along with them, constantly over the course of September 11 and the days that followed.

As private citizens defined the disaster in their own terms, they did so not in isolation but in negotiation and deliberation with one another, with officials, and with their broader environment. This includes the purposes and importance of certain rules. And they did so not only once but over and over again. Later, we discuss how playing with the rules in this circumstance led to successes rather than an abundance of mistakes and mishaps from which the emergent operation could not recover.

Rule Breaking with Vigilance

The Coast Guard had a hand, too, in breaking rules, including its own rules for maritime operations. Indeed, in Day's Coast Guard oral history of the event, he jokes, "I broke more rules than probably I've enforced in my whole Coast Guard career." Although probably not true, his comment embodies many of the stories we heard about the event: just because rules were broken does not mean that there was a lack of order, organization, or concern for safety. Rules were broken, but new rules were made, and then these rules were broken throughout the duration of the maritime activities with equal skill by volunteers and officials.

It is critical to our analysis to note that rules were not being broken heedlessly or recklessly. They were being thoughtfully disregarded, even in the desperation of those first hours when people just wanted to do *anything*. We call this *rule breaking with vigilance*. Everyone broke the rules, but they broke them gracefully, with sensitivity for consequences and with a sure-footed sweep through a potential minefield of possible mistakes and accidents. What guided these maneuvers? What made it possible to suspend rules for safe operations, rules that had been carefully developed over years of experience, social consensus, and technical judgment?

To begin with, while the rules themselves were broken, their purpose was not. To put this another way: the law was being broken, but the spirit of the law was not. The sailors and waterfront workers held to the fundamental purpose of the rules, which was safety, a concept expanded in this case to include transporting people from the dust, ash, and smoke of Lower Manhattan. Day recalled a point when one of the pilots suggested that removing the large boats from North Cove Yacht Harbor would give the operators more room to maneuver. So they moved them, towing a string of multi-million-dollar yachts up the river. Day did not even really know what would have been involved in getting permission to take possession of those vessels. How does a Coast Guard officer impound someone's private property? Perhaps with authorization from the attorney general? But the decision made sense to him: "There were a lot of little decisions like that, and that's the call. And it was easier as time went on to do stuff like that. Once I had rationalized it as a way [of] why I was doing things, it kind of became clear to me why I was doing things." This is a wonderful example of making sense while in the process of doing.

Remember how McGovern had indicated that the *New York* would need some Coasties? One officer even gave a harbor pilot a Coast Guard ball cap and lifejacket and said to him, "You're in the Coast Guard if anybody asks. Go right in there." Impersonating a Coast Guard officer is a federal offense! But, maybe not, under the circumstances. Complex, and new, organizational bonds were emerging. Organizations, and vessels, were extending their tasks and expanding the personnel who were helping achieve those goals. Extending Coast Guard legitimacy to these individuals, at least during this portion of the improvised operation, was part of that effort.

Day described how "fire engines were starting to run out of fuel along the way, and someone came up to me and said, 'Hey, you know, if we line up some tugboats here, we can—they have diesel—we can fuel these fire trucks.'" It made sense, but he could not exactly write up a purchase order for fuel on the spot. Indeed, the fiscal part was where he started to get a little nervous. He emphasized that the transfer of fuel was not for the Coast Guard to decide—after all, he did not want to be on the hook for the cost of a million gallons of fuel—so others worked it out among themselves. From those we talked with about the incident, the sentiment emerged that this was a question of which was worse, stealing fuel or losing lives because of trucks running low on fuel. Breaking rules to prevent the loss of life ultimately won out in this impromptu risk calculation.

Day recalled one person who protested the refueling efforts at the site without a permit: "'A permit?' 'No,' I said, 'it's authorized under the Coast Guard in relation to an emergency.' . . . I said, 'No, no, it's a hierarchy of laws doctrine that we all can do this.' . . . He was like, 'All right,' and he walked off, and the chief was like, 'What's this hierarchy of law doctrine?' and I go, 'I don't know.'" Day had made it up, but it worked.

Other rules and norms were suspended. Responders chopped down fences. They tied their boats to other people's trees. Accepted concepts of private property were modified. Of course, not everyone agreed that the rules should be broken. We heard numerous reports of Battery Park employees (or perhaps it was just one or two in the vicinity of a lot of other participants) who became angry regarding some of the steps being taken to facilitate the boat evacuation, especially the fence cutting. One person we talked with recalled a Parks Department employee saying, "You cut a railing and [we are going to do] this and that." They cut the fence down anyway.

Another participant in the evacuation recounted the resistance of a Battery Park employee when captains tried to tie up their boats to trees along the shore. "He tells me, 'You can't tie up to those trees. You're going to ruin those trees.' And I don't want to repeat what I said to him, but he almost went in the water. We didn't see him again. We got rid of him." We heard many reports of male Battery Park employees contesting the tying of boats to trees. The sheer number of sites in which confrontations occurred, and the occasional reference to male and female Parks Department employees, suggests that more than one person from the Parks Department or Battery City Park (both entities were cited) was involved. The difference in perspective seems to have been organizational rather than specific to a particular individual (although there were some exceptions, too, as noted in Steven Whitrack's interaction with a Parks Department employee in Chapter 3). A mariner recalled, "She said, 'You can't do that!' and it's, like, 'Well, tough, are you going to arrest me?'" Several years later, the recollection still generated a sense of bitterness over the clear difference in perspectives.

Another person in a different area of Lower Manhattan recalled that as people were embarking on the boats and as supplies were being taken off, a member of the Coast Guard lamented that life would be easier if the railing were not there: "[My thought was] 'Well, get a cutting torch and cut the railing off.' '[But] it's not our property,' [the Coast Guard officer replied back]. You know, people kept forgetting that . . . [a] couple of buildings fell down, and you could go outside the box, you know, and when they did, it was amazing."

In some areas, rules began to tighten rather than relax. As we noted earlier, Liberty State Park, where Liberty Landing and the marina were located, was one site for disembarking the evacuees. According to one person we spoke with, after a while, Liberty State Park seemed to be "under martial law, pretty much. You couldn't get in and out of there unless you belonged there, and [the Park Police] were pretty adamant about it." If you left the park without identification, good luck getting back in. We heard a story of a woman and her baby who were staying with someone on a yacht in the marina. When the woman's family was finally able to come get her two days later, the Park Police would not let them in, nor would the Park Police drive the woman and her baby out to meet her family, since the squad car lacked a child safety seat. Instead, she had to walk more than a mile with her child to meet them. Another person we talked with described the Park Police in a less-than-kind manner: "I just thought they were weenies."

To attribute this perspective solely to some personality characteristic of certain members of the Park Police misses the mark. The organizational culture of the Park Police differed markedly from that of the marina residents and many of the maritime organizations within the park. The Park Police were rule-focused, and they had a security and law enforcement frame that was, quite reasonably, influenced by the hostile nature of the attacks. Those working the checkpoints had less authority to make independent decisions; it became less and less of an option for these employees to deviate from their instructions the more time passed from the moment the disaster began. Likewise, the Manhattan-based Parks Department employees operated under a different organizational framework that clearly did not adjust to the emergent environment. They were less able to accommodate a situation where the fingers of improvisation could slip in between the fingers of planning to ensure a firm grasp on the response.

Nor were security checks limited to Liberty State Park. At Sea Streak in New Jersey, for example, officials began asking for identification as those fleeing Manhattan disembarked, and some without identification were questioned. Mariners on at least one vessel began checking backpacks, without any guidance or directive to do so, but out of a sense that the city was under attack. Some of the crew remembered that terrorists sometimes engage in secondary strikes on rescue workers coming in to help. One thought a police or security officer on board might have initially suggested the procedure, but no one was entirely sure. The fact that these particular mariners had seen, quite clearly, the second plane hit, and that many of them had military training, likely influenced their decision to be mindful of security in ways that did not trouble other mariners. They had heard rumors of more planes coming in. Yet these new procedures slowed the disembarkation of evacuees and thus delayed the turnaround of the boats back to Manhattan.

Decision Guidance

None of this is to say that rule-breaking decisions came easily or that everybody could tolerate any relaxation of normal procedures. Day said, "You look and wonder if others are crossing the line. You wonder if you have crossed it." Day was apprehensive, too, and noted that at the time, some of those boats full of evacuees seemed particularly low in the water. His anxiety was shared by the boat operators themselves.

Shared anxiety seemed to be key in the collective balancing of risks. Day was imagining the worst-case scenarios—adding a capsized ferry to the list of challenges that day. But he was also imagining the consequences of inaction. With each decision, he asked himself whether he was doing what was right and whether he felt comfortable making the authorization. He knew the command culture. He knew he would have the support of his superiors provided he did what was reasonable. He knew there was trust. And they knew he was not going to do "something totally wacky." It turned out that nothing "wacky" came up that he had to say no to. But at the same time, he understood that there were limits to the rule breaking.

If Day could fall back on command culture and the trust of his supervisors, what were the mariners relying on to guide their decisions? Mariners, as we have seen, already have institutional resources for evaluating risk. Given the urgency of the situation, mariners substituted their own judgment over the Coast Guard's rules in, for instance, determining for themselves how much of a safety margin in the boarding of passengers was adequate. Boat operators stressed to the research team that such decisions were not made recklessly but based on their assessment of each boat's performance, the distribution of additional weight, and the demands of the immediate crisis. Take, for example, this account from one of those mariners:

> We were supposed to put eighty passengers; we were putting about ninety to ninety-five . . . but I kept basically everybody on the lower deck for the stability. . . . I [knew] what I was doing; the deckhands knew what they were doing. We were going against the Rules of the Road . . . but we were still keeping within a safe manner. . . . They were just stretching the limitations. . . . One more thing I did—I made sure everybody had life jackets on, just in case.

The Coast Guard eventually assisted in this rule breaking, sending inspectors like Concepcion to the waterfront to perform spot checks to ensure vessels were not dangerously overloaded. Concepcion knew the capacities of these craft, but he also knew many of the captains. Some dinner cruise boats could transport more than three hundred people but needed crew in addition to their captains. The problem was that crew members were not able to get to the area. The captains proposed using state troopers to handle the lines, and Day said okay: "It seemed that some of those things made sense at the time. . . . I'd worry about it later. . . . It would just come down to, you know, defending my decision at that point." He did not want a decision he had made to result in a worse situation, but it made sense, and he worked for a commanding officer whose leadership style was "do what makes sense."

Certainly the usual margin of safety adjusted in this event with respect to capacity and to navigation. For example, in some areas around the harbor, the dust and ash were so thick that visibility was reduced to zero, yet some operators chose to enter the breakwater at North Cove on radar alone. One ferry skipper who shuttled evacuees to New Jersey's Liberty Landing called the process "a lot of blood, sweat, and tears." Maneuvering into North Cove was a tense job. His boat was fiberglass—"cement and fiberglass don't go good together." He could not see with the thick smoke and dust. A normal day of fog or rain would have been like Christmas compared to this, he said. Offshore, the shoreline's image on the radar screen looks a lot like its outline on a chart. But close range distorts the perspective because the shoreline is increasingly under the radar antenna, not in front of it. The image becomes blurred, as the shoreline gets confused with the electronic echoes of buildings. The radar picture, up close, is more like an impressionistic painting than a clear, detailed photograph. There is an art to using radar and interpreting the electronic landscape, and this skipper relied on that art to feel his way into North Cove. He remembered thinking, "Radar, don't fail me now!"

Taking those kinds of risks was still not easy for these mariners—what if there was an accident? Mariners are always alert to the possibility of litigation and being called up to justify their actions at a board of investigation. Military personnel are as concerned as other mariners about boards of inquiry. The mariners that day had broader constituencies in mind when thinking about the consequences of their actions. For Day, he was aware of two larger constituencies for his judgments: first, his superiors, and then a wider second sphere. Day described what we like to call the Mike Wallace barometer, based on Day's reference to the journalist who served for nearly forty years as a full-time correspondent on the television news magazine *60 Minutes*:

> I was thinking . . . [if] I was on *60 Minutes* and they were interviewing me, would it make sense . . . for me to be answering to Mike Wallace that I . . . slowed [the evacuation] down because of a rule that the Coast Guard enacted and it just didn't make sense? . . . I mean, I was concerned, and I thought about it. It was definitely one of these moments that I was very, very apprehensive, just because some of the boats were really low. . . . I kind of came to this: it's okay, you know; it's the right thing, and I feel better saying yes. I authorized [it], and here's why.

In fact, a few days later, in the wee hours of the morning, Day was finally able to touch base with his commanding officer to update him on the activities, an update that was in part a confessional on his various transgressions: "I remember him saying it—'Keep doing it; you're doing great.'" But in those sudden moments when questions arose, with the dodgy communications and the intense need for quick decisions, Day could not bump questions up the chain of command. He was the one in front; the Coast Guard's role and the resources that it provided rested on his discretion. Day's Mike Wallace barometer served as his guide: how would his action or inaction look to a larger public?

Ultimately, Day's decisions, and those of others we interviewed, were moral considerations. They wanted to do what was right, to rope together ideas, principles, and existing procedures into a bridge for crossing over from doubt and ambiguity to certainty. Both making and breaking rules have moral implications. Some principles of moral philosophy provide insight for bounding good or bad rule breaking. The philosopher Sissela Bok (1978) has studied a special kind of rule breaking: lying. Her work examines when it is acceptable to lie, eventually concluding that lying is only rarely acceptable, only in situations where the lie could stand the scrutiny of comparison to higher ethical imperatives. This same reasoning on the ethics of rule breaking emerged collectively among the mariners on 9/11.

The aforementioned examples from the Parks Department serve as a reminder that some organizations are more willing than others to set aside rules and demonstrate flexibility in an emergent environment. Day explained how the Coast Guard enacts a remarkably high level of freedom and latitude that works to empower those at the lowest levels to make decisions. With fewer personnel compared to some other branches of the military, bureaucracy is consequently less of a problem. But Day argued it was more than that: "I think our culture is one of you can be two hundred miles offshore, and you really can't ask how to go about [something]. So, it's do the mission, you know, if there's people. You know, improvising. It's just, I think, a way of life in the Coast Guard." Indeed, the U.S. Coast Guard was one of the few federal organizations that escaped Hurricane Katrina without scathing criticism, largely because personnel felt empowered to make sensible decisions without always getting permission—a key feature of organizational reliability (Mallak 1998).

But that culture of flexibility and adaptation is arguably part of maritime culture, more generally. As one tugboat captain put it:

> Because it's the nature of the business, you deal with things under pressure. You deal with Mother Nature, you know, and you have to deal with what she deals. And you can't say, "I want to go home because it's raining out." That's . . . part of the

> job. And you deal with what happens, you know. You want to keep the boat safe and the crew safe, then you get the job done. And the job that day was moving the people out of the Battery.

One person we talked with recalled a mayor's emergency management meeting sometime on Tuesday. Someone in the mayor's office said something along the lines of "Everybody here has something that they can do for us and give us. I don't know what it is. So we would like you to come up with viable plans within the next three hours and come back here for the next meeting with anything you think can help the situation." Fewer than twelve hours after the attacks, officials were willing to listen to suggestions. They approached the overwhelming sets of problems that lay in front of them with an openness—at least for those at that meeting—to the range of skills, knowledge, and ideas others could bring to bear on the problems. Was it just because the event was so big and so complex? Was it because those close to the response in those early hours quickly developed a connection with those around them? Whatever the cause, framing the response as one where others have untapped and unknown resources to offer helped shape the activities in a way that welcomed the input of those beyond the usual suspects.

As time went on, the window of adaptability closed. We heard examples of turf battles and bureaucratic concerns reemerging as barriers in various circumstances. The same person who noticed the openness in the first meeting provided this account about how the flexible environment began to change:

> They could not think outside of the box. Almost every city worker I came into contact with could not think outside of the box. They had to talk to their boss. They had to talk to their boss's boss. They were all trapped by their department regulations.

Another mariner described it this way:

> I truly think that rules and procedures were the biggest hindrance of the whole thing. . . . I mean, who doesn't like a fire-

> man? Wonderful guys. [And] the policemen. . . . [But] we walked up to a policeman, and we needed stuff, and he goes, "I'll check with my, my sergeant the next time he comes around," and, like, for Christ's sakes, can't you think? You know, we need something done now, [and] you're going to wait 'til your sergeant shows up again?

Another person we talked with commented that "the system of response that we have now is that it does not, for national disasters, lead to [an] overwhelming response. It's all about liability and not about life safety."

What about other examples of the response that involved setting aside rules that actually led to dire outcomes? On the morning of September 11, some firefighters self-deployed and left their firehouses in spite of instructions to remain. Others, in the days after, ignored orders to leave Ground Zero after officials tried to limit the number of people at the site, instead staying out of a genuine compulsion to find their fallen colleagues. Still others set aside personal protective equipment because of its impediment to vision and communication, putting their health at risk. We would argue that many of these decisions did not involve the same level of vigilant deliberation that we saw with the otherwise successful rule-breaking situations we heard about. The mariners who were rule breaking successfully were doing so in domains in which they had expertise. Had they engaged in rule breaking in other domains, perhaps more missteps would have emerged.

One of the popular conceptions of leadership, virtually a stereotype, is of the single individual making all decisions, a kind of all-knowing figure who gathers information from various sources, processes it through unique personal capacities, and tells others what to do. But plenty of evidence, especially in this event, shows that this trait is not at all what leadership is about. Instead, leadership is in many ways a group activity, a communal product. Even though ultimately one person may bear responsibility for the outcome of a decision, the prudent leader is alert for contrary opinions and monitors the effect of orders on the various constituencies, which include subordinates, superiors, and the general public who will experience the effects of those orders (Denhardt, Denhardt, and Aristigueta 2009).

Throughout 9/11, people lacked experience with the multiple demands and often had no guideposts or prior decisions to use as benchmarks for decisions to be made. Day was concerned about getting things right, but how could he know whether he was, in the absence of any precedent for the job he was doing? The collective judgment and wisdom of Day's colleagues substituted for experience: "I think if I was off base, they would have told me. I think they would have; I know they would have. I trust that they would have said, 'I don't know Mike, rethink that one.' And they don't work for me; I work with them."

Yet Day's choice to bring along Chief Wilson suggests that he factored in this concern, bringing someone who would not be dissuaded by hierarchy or rank from saying what he thought: "I knew his personality, and I wanted that personality there. I didn't want to be getting in arguments . . . but I still knew him well enough that he would speak truth to power, if you will. I trusted his judgment." Someone else might be intimidated by rank. Someone else may not call out something that did not make sense. This person would: "It was a personality call." They helped each other learn how far the rules could be stretched or what new rules could be made.

In these instances, we saw the maritime community navigating contradictions. They knew what they wanted to do: they wanted to transport people, but they knew doing so would require a lot of liberties and actions that they normally would not take. They knew they were pushing boundaries, making their own decisions about safety, forming new organizational relationships, and acting independently from their superiors. Sometimes they were nervous, hoping they would not cause a worse problem. And then they caught their breath, rethinking what they were doing. Remember what someone said: "Let's not end up with a bigger situation that doesn't have to be."

Should people just go around breaking rules? Not ordinarily. But most of the time, as far as we could tell after one hundred interviews, the migration of decision-making authority (Weick, Sutcliff, and Obstfeld 1999), including bending and breaking rules and making others, was helpful when two conditions were met: (1) the disregard of rules was thoughtful, and (2) the underlying purpose of what the rules were set out to achieve was preserved. Often there was a third element: deliberation with others. In many instances, people talked it out. They sought guidance and reinforcement, checking themselves against others' interpretations of the situation, as Day did. They did not rely on a single judgment but followed through on the maritime injunction against trusting any single piece of information. There was so much to be done that there was plenty of room for people to step in. But they did so with a sense of self-reflection and by attuning themselves to their environments. They watched for opportunities, for gaps to fill with needed work.

5 /

Blending Art and Science, or Mindful Muddling

Toward a New Concept of Disaster Management

During the waterborne evacuation of Manhattan, many emergent, unscripted activities came together to form a successful response to an unprecedented disaster. All things considered, it went relatively smoothly. Mariners, who saw in a sudden and surprising event possibilities for helping that allowed them to deploy their skills and equipment, made up the core of this component of this response. The habits and experiences of the mariners, and their exposure to uncertainty and doubt in their normal work, pushed them to make sense of the situation first through their identities and later through their actions. But mariners were not alone in this operation. Other people who did not necessarily see their ordinary skills in terms of emergency management nevertheless found ways to help by using their skills, whether as logistics coordinators or simply as people who could serve food. More than anything else, the waterborne evacuation of Manhattan suggests that it is possible to pull off significant disaster management work with no preplanning.

The success of this operation gives us the opportunity to glean some insight into what makes effective improvised activities possible. One requirement is a strong local network and sense of community. Over and over again, we were told that everyone in the harbor knew one another, that it was a close-knit community, and that they were familiar with the strengths and weaknesses and capacities of others in the network. Moreover, several different "communities" were involved, but they were scattered around the harbor. There was no effort to join them into a unified system, because there was no unified event. A lot was happening all at once, with challenges tackled by whoever was closest to the action. The Coast Guard, for instance, made the wise decision to use its resources of legitimacy and federal power in support of an existing maritime community rather than attempt to impose some external meaning on what was happening.

A second resource that contributed to the success of the operation was responders' deep level of local environmental knowledge. The boat operators, and especially the Sandy Hook pilots, had a vast knowledge of the local area; pilots held detailed charts of the waterways in their heads after years of study and thousands of trips on ships of all sorts. This knowledge went beyond knowing the water and land; members of the harbor community shared a baseline familiarity with how boats and ships fit into the dynamic world of maritime commerce. They knew how technology and environment went together to enable moving goods and people—their principal job here. Whether in a forest, on a mountain, in the desert, or in a city, effective disaster management needs this level of environmental knowledge.

A third factor was a willingness to be flexible with existing rules and procedures while acknowledging that new rules might be required. Using tugs as passenger ferries, loading boats beyond capacity, getting fuel where it was available, and taking property with or without permission made it possible to match resources with demands. Cutting down fences is perhaps the best metaphor: when Paul Amico and others cut down iron and wooden fences, they were literally and figuratively clearing the way for new actions. At the same time, no one was taking promiscuous risks. As best we are able to determine, risky actions were taken with a calculated awareness of the consequences of breaking the rules versus the urgency of the situation. The reasons for rules were not forgotten, but the boundaries were stretched.

These observations point to a particular concept of disaster management: one that sees plans as tools, not scripts; one that is respectful of new and unexpected resources; one tilted more toward effectiveness than efficiency; and one that liberates the improvisational and creative powers of an affected population to make meaning and solve problems, supported and assisted by external resources and advice. While this conclusion may sound intuitive, or even obvious, it challenges the concept of disaster management that currently prevails in the United States and, indeed, much of the rest of the world.

Two schools of thought have emerged regarding the proper, most effective, organizational design for managing a disaster. Both of these schools—each focused around the concept of the Incident Command System (ICS)—depend on the imposition of external systems on communities in crisis. One ICS approach emphasizes centralized, hierarchical control, in which a commander oversees a number of sections (planning, operations, logistics, and finance) that carry out all the emergency operations. The concept has long been criticized by scholars who believe that it bureaucratizes what is essentially an open system and therefore is flawed at a deep theoretical level (Buck, Trainor, and Aguirre 2006). Another interpretation of how the system works is suggested by Donald Moynihan (2009), who argues that the "command" aspect of ICS may be overstated and prefers to construe it as a system for network management. Both views of ICS assert, nevertheless, that it tries to bring within its circumference the various responding organizations.

A different theory of disaster management is suggested by the boat evacuation: management based on coordination by noninterference, or a redefinition of disaster activities as allied modules rather than a holistic connected network. This approach draws on a different concept of disaster management, commonly called the Emergent Human Resources Model (EHRM; Neal and Phillips 1995), which favors problem solving by people close to the action. In the largest and most surprising events, coordination and leadership as dreamed of by officials is not possible. Situations change quickly, and it is nearly impossible for the full range of responders to know who might enter the disaster-response milieu. We found in our examination of the boat evacuation evidence that full situational awareness was never achieved. What we saw was the harnessing of the principles of pooled interdependence (Daft 2004)—people working together on tasks that did not necessarily connect. The EHRM model describes the involvement of disaster response participants as "flexible, malleable, loosely coupled, organizational configurations" (Neal and Phillips 1995: 327). We take that idea one step further. In practice, there is no single disaster that everyone is tackling. Those loosely coupled configurations are not responding to a *single* event. Rather, they are allied modules responding to their individual pieces of the disaster. This approach is what in the end makes an improvised disaster response manageable. It is not improvising the entire response; it is improvising the aspect of the response that you and those within that module of the emergent network can handle. An effective disaster response, we argue, can be built from diffuse sensemaking.

Current Models for Disaster Management

Before we progress too far in this argument, it is worthwhile to take a closer look at the current U.S. system for disaster management, including ICS. Many of the shifts in operating concepts and policies in disaster response have emerged directly from frustration with previous policies. The Federal Emergency Management Agency (FEMA), for example, owes its existence to such frustration. President Jimmy Carter established FEMA in 1979 in response to a growing concern that disaster management was fragmented across too many agencies (Knowles 2011). By creating the agency, the Carter administration hoped to gather in one organization a number of functions that had previously been dispersed through many offices. Yet the creation of FEMA did little to change the fundamental concept of American disaster response—that is, to coordinate the activities of other organizations' personnel and equipment. Thus, every time that FEMA mobilizes, it builds a new organization from scratch. Even with plans, training, and exercises, each disaster yields a fresh response system.

This pattern repeats itself at federal, state, and local levels. All emergency management agencies in the United States are essentially coordinating agencies, with few resources of their own. Their jobs are to pull together police, fire, medical, and humanitarian services, whether those organizations are government entities or private service organizations (for instance, the American Red Cross). But again, the actual emergency management organization, when activated, is temporary and transient. The stories, myths, and lore that convey learning (Weick 1995) in a permanent organization are sometimes lost in the constant disbanding.

Aside from the lack of institutional memory, U.S. emergency officials have long struggled with the challenges of integrating emergency response agencies across jurisdictions. Some cities and counties have mutual-aid agreements to share resources if one place is overwhelmed; others have memoranda of understanding to bring in the private or nonprofit sectors. A sequence of destructive wildfires in the early 1970s in California highlighted some particular challenges of coordination, including different radio procedures (such as variations on the famous "10" codes), names for equipment, and duplications of resource requests. Fire trucks notoriously passed one another going to fires in opposite directions (Irwin 1989). Afterward, a consortium of public officials, emergency experts, and academics from the business and management sciences examined what went wrong from organizational and technical perspectives. They identified a "lack of a common organization," "poor on-scene and inter-agency communications," and poor planning and information management, among other deficiencies (Irwin 1989: 86).

ICS design principles were developed to meet these deficiencies, as discussed further by Robert Irwin (1989), to include a capacity for operations involving single or multiple jurisdictions or agencies, applications for various types of emergencies, and adaptability to shifts in technology. Over time, in slightly different manifestations, ICS became accepted and even, in some instances, mandated as the approved system to use in emergency management. For example, ICS has long been mandatory in hazardous materials responses. More recently, the National Incident Management System (NIMS), developed after the attack on the World Trade Center, requires the use of ICS in all disasters. As a consequence, emergency responders nationwide must now complete a number of ICS training segments to be in compliance with federal guidelines. One notable consequence of the NIMS concept is the expansion of the definition of a "responder"; school administrators, utility company workers, and hospital workers now must know something about ICS, since any or all of them may someday be involved with emergency activities.

System designers and users hope that the organizational structure prescribed by ICS can provide a platform for agile and disciplined responses (Harrald 2006) and serve as a substitute for the lack of a full-time, national, fully integrated, and permanent emergency management system. In the U.S. system, ICS is intended to provide the equivalent organizational features of common understandings and shared expectations that one sees in permanent organizations. At the same time, ICS is controversial. While many emergency managers love it, many scholars assert that its benefits are wanting or ambiguous. On the most basic level, critics claim that ICS fails to take into account perspectives beyond the responder community. Moreover, some say the system is needlessly complex: the full articulation of the system used in the United States requires thousands of pages of text and many training courses to explain the concept and functional details.

Other critiques surround the extent to which ICS takes into account the specific characteristics of an event and the participants in the response. Dick Buck, Joseph Trainor, and Benigno Aguirre (2006) have studied a dozen emergency response incidents and concluded that no sweeping generalization about the usefulness of ICS was possible. They argue that its effectiveness in one disaster or another owed much to the nature of the disaster itself and the characteristics of the interorganizational network that developed to confront the crisis. In some situations, ICS has proven to be an effective management protocol; in others, it has been less effective. For example, they have found ICS to be most effective in situations where a relatively small number of well-socialized organizations have worked and trained intensively together before the event. ICS is also effective when dealing with familiar types of emergencies with well-established and practiced response strategies and a low-level of emergent activity. They have found it to be less effective when dealing with novel, surprising, or outsize events.

Yet another observation about ICS is that, while its rhetoric and training stress preestablished roles and procedures, it actually relies extensively on consensus, coordination, and negotiation. In our own research on 9/11, for example, we found that considerable negotiation for resources was taking place outside the formalized approval system for logistical requests (E-Team) at the city's reestablished emergency operations center at Pier 92. Instead of solely relying on the E-Team software and newly established procedures, in many cases people took the initiative to secure the supplies needed, informed those who would eventually receive their "requests," and then entered the request into the software system for formal approval and tracking purposes. The initiative of those many individual efforts gave the response the appearance of a well-oiled hierarchical system, even when it relied on decentralized initiative. Among the hallmark characteristics of ICS are *joint planning*, an *integrated action plan*, and *management by objectives*. All of these require the participation and agreement of the relevant agencies, all of which are supposed to be represented at incident command briefings. While ICS supplies a hierarchical communications and reporting framework, consensus, coordination, and negotiation are not individual or organizational capacities that can be commanded.

Simply stated, social scientists have repeatedly found that ICS does not deliver the universal coordination promised in its design criteria (e.g., Buck, Trainor, and Aguirre 2006). Most scholars in the disaster field, including us, find the ICS concept to be unsatisfactory. The approach views organizations as mechanistic structures and relies on principles that find little application even in ordinary business or commercial organizations, let alone in the management of disasters (Sylves 2007; Waugh and Sylves 2002). Social scientists object to ICS's command-and-control aspect, arguing that, in disaster, much activity goes beyond command approaches (Drabek and McEntire 2002).

First responders and emergency managers, in contrast, tend to be strong supporters of ICS, arguing that its failures might be traced to lack of training in its principles or faulty application during a response. Their most compelling rejoinder, however, is to ask what the alternative is. It is a legitimate question for organizations increasingly held legally and morally accountable for their responses. We suggest that ICS—or something like it—might be fine for those organizations that can be captured *reasonably* within its structure. But we also suggest that some organizations cannot be forced into that model. A reconsideration of how an ICS-like system could better function as an open system in concert with an EHRM model as well as a better appreciation of the lessons from the waterborne evacuation offer a promising new direction.

Aside from the waterborne evacuation of Manhattan, we have studied such large-scale events as the 2004 Indian Ocean tsunami, the 2010 Haiti earthquake, and the 2011 Tohoku earthquake and tsunami in Japan. In each of these cases, we found no evidence that ICS did or even could have structured the response to such complex crises that presented strong elements of surprise and novelty across large geographic areas. In each case, the response needs were conflicted, contradictory, or ambiguous. This quality of ambiguity is important. As James Mitchell (1999) has noted, "Ambiguity connotes circumstances of indecision—where customary guides to choice are missing, non-functioning, undependable or so deeply conflicted that decision making is effectively paralyzed."

Ambiguity is likely to be present in almost any large-scale emergency—precisely the kind of event where ICS benefits are most hoped for. At the same time, this ambiguity makes the imposition of an ICS response structurally impossible. This dilemma is one with which scholars have struggled for decades: the need to bring conceptual and therefore functional clarity to a fluid and dispersed event while simultaneously enabling flexible and responsive action. Karl Weick (1987: 124, citing Perrow 1977), for example, refers to the need to centralize cultural and normative expectations before you can decentralize: "Before you can decentralize, you first have to centralize so that people are socialized to use similar decision premises." Participants need to have some unifying or familiar touchstones or reference points for their behavior. Louise Comfort (1999: 4) thinks that a sense of "shared risk," enabled through improved mapping, computing, and communications technologies, would foster flexibility and coordination. We have seen in the waterborne evacuation, however, that these important characteristics can also emerge from existing social constructs.

ICS is a system that must be *created* by the participants themselves. Moreover, as we have noted, it must be established anew each time. It is not an automated system but rather one that depends heavily on the participants themselves and their capacities for observation, interpretation, and, indeed, creativity. The participants are the ones who have to define the extent of the incident, for example, and determine whether single, unified, or area command is appropriate. Single, unified, or area commanders negotiate the extent of the disaster and the corresponding extent of the ICS that they build. Participants may occupy organizational squares with standard names, but each participant must fill the square with his or her individual capacities and interpretations.

In hindsight, some might describe Mike Day as the incident commander of the boat evacuation, but Day certainly did not command the entire 9/11 response. And one would be hard-pressed to describe the way Day approached his activities as that of a true commander, or "in charge." In hindsight, one might be able to put various individuals into various boxes, but they did not see themselves necessarily occupying a box at the time and felt a fluidity to move to other roles as deemed necessary; additionally, some boxes that would otherwise be prescribed as part of an ICS system were not—or not always—attended to with someone seemingly in charge. Some might say that bringing one's own qualities and interpretations to the squares of ICS is how the system is meant to function, but a system that dictates organization, command, and accountability under prescribed designations without actually anticipating or accounting for how activities operate is not a system that reflects reality: it is a fetish object for people in organizations who have to demonstrate accountability and why something worked or failed.

This is not to argue that ICS is wholly unworkable, only that establishing the necessary communicative capacity takes time and practice and that such practice has to be on a scale commensurate with the expected event. In other words, planning for an event the size of Hurricane Katrina would likely require many years of extensive and meaningful exercise to build up the required levels of familiarity and trust between the relevant organizations. This sort of relationship building would have to occur on an ongoing basis, as organizations and organizational actors change. An organization that is prepared for a large event can, however, easily scale down for smaller events. As an example, the crash of American Airlines Flight 587 into the Rockaway Beach section of Queens, New York, on November 12, 2001, was handled swiftly by responders, as we personally observed in the Emergency Operations Center (EOC). Certainly it helped that the EOC was already activated and that responders of every sort were plentiful in the city. Compared to September 11, the crash of a commercial airplane into a residential neighborhood seemed comparatively easy to absorb. Of course, prior to September 11, that same event would have been among the worst-case scenarios for New York's emergency managers. An enhanced imagination for disaster, coupled with the response "practice" tragically initiated by the World Trade Center attacks, had bolstered New York's response repertoire and capacity.

It is not our contention that there is no place for ICS in the management of emergencies, even complex emergencies. But its application will suffer if those in charge of its implementation fail to understand the difference between emergencies, disasters, and catastrophes (Quarantelli 2006; Wachtendorf, Brown, and Holguin-Veras 2013). Just as a disaster is not merely a large emergency (Auf der Heide 1989; Quarantelli 2000), a command system for a catastrophic event cannot simply be an expansion of a system designed for a smaller scale. This idea is contrary to one of the founding principles of ICS, which is that it can be expanded to cover an incident of any size or complexity. But simply adding modules to cover the imagined larger scope is not enough. If we argue that emergencies, disasters, and catastrophes are qualitatively different events, then our conception of an appropriate management system must be similarly enriched. Even the term *incident* implies something small. For all the criticism lodged against Mayor Ray Nagin of New Orleans and other local officials regarding their response to Hurricane Katrina, Nagin's statement that federal officials were "thinking small" (CNN 2005c) is apropos.

The disaster management system is not an automated entity that accomplishes its functions as soon as it is implemented. Rather, it relies on people who are empowered to enact the system. John Kelly and David Stark (2002) have researched the recovery of firms that were affected by the attacks on the World Trade Center. These firms lost many employees, documents, files, and workspaces. Following their interviews and focus groups with some of the workers in these businesses, Kelly and Stark observe, "No one said, 'Our technology saved us' or 'Our plan really worked.' To [every last] person, they said, 'It was people'" (1524).

This human aspect of crisis response is subordinated in ICS doctrine. Weick (1987), for instance, has made the point that what makes air-traffic control a reliable system is that the controllers themselves create the system by varying such parameters as aircraft altitude, speed, and distance. With their backgrounds (often ex-military), training, and socialization, they bring the necessary comprehension. At the same time, their freedom to act provides the necessary agility to make the "system" work from day to day.

ICS, implicitly, is an organization for leadership—if there is an incident commander, he or she must be leading, somehow. At the same time, principles of leadership and management in modern organizations that operate in dynamic environments recognize the need for flexibility and autonomous action by people who are closest to the work. Leadership becomes not a command activity but a facilitating activity. Clearly, no organization could continue to function if everyone operated on his or her own without regard for how individual actions affected colleagues or the organization; flexibility and autonomy do not necessarily mean that people are working utterly independently. Organizations implement various strategies that allow for flexibility but also maintain focus on organizational goals, becoming skillful at shifting between various structural forms that maintain order while allowing workers to meet shifting needs (Daft 2004).

The various responder organizations must act in concert with one another, yet they must do so with independence and autonomy, especially in the early stages of a disaster when information is scarce and conflicting. Even having access to the same information does not guarantee a smooth response, since interpretations are likely to differ at various operational scales. The dilemma is difficult. Given that ICS seems to work well in emergencies that are contained, familiar, or localized, the real challenge for emergency management is in large, regional events that involve many jurisdictions simultaneously and demand a variety of response functions. A single organizational structure for such a situation is probably not possible.

Lessons from 9/11

What can we learn from the waterborne evacuation that helps us reconcile these issues?

Maritime experience provides a metaphor and a guide for a different concept of emergency management. There is barely a navigation "system" in the United States. Existing Vessel Traffic Services (VTS) have little directive authority. Although VTS will warn a ship steering into danger, no one tells the vast population of boats and ships where to go, as in aviation. In conditions of distrust and conflict—when fundamental relationships between the main partners in navigation are strained, when equipment that designers and regulators tout as the salvation of navigation is itself prone to error or not to be trusted, when danger can surprise at any time, when the venerated Rules of the Road provide only approximate guides—mariners get used to muddling through (Lindblom 1959). To whatever extent this "system" works, it works because mariners make it work, collectively, in their day-to-day practices of steering and watching. Mariners work in situations that are basically out of control, where the only control is what they themselves inscribe over the water.

Such a system is a metaphor for managing in a disaster. In spite of any number of efforts, no one has been able to inscribe a comprehensive system over the uncontrolled milieu of disaster, any more than the mariners' Rules of the Road provide a conclusive guide to action in any given navigational situation. While the shifting and turbulent waters are real and disaster is the product of a collective imagination, the consequence is the same: people make their own protean institutions whenever they venture into demanding and unfamiliar environments. The mariners of 9/11 grafted their *familiar* uncontrolled environment, with its specific coping methods, onto an *unfamiliar* uncontrolled environment in a way that allowed them to keep up familiar habits of thinking and interacting. Captains drove boats, pilots directed traffic, they called on the radio, or they just kept out of one another's way. They tried to be ready for anything; they tried to keep alert for the unexpected.

This skill of grappling with ambiguity is, of course, not limited to mariners. Social service agencies are generally staffed by people who are good with working with clients who are in deep trouble, who are used to scrounging for resources and patching together networks of care. They seem to adapt quickly in disaster, absorbing, as it were, the disaster into their usual organizational cognition. In her work examining domestic violence shelters after Hurricanes Katrina and Rita, Bethany Brown and colleagues (Brown 2009, 2012; Brown, Jenkins, and Wachtendorf 2010; Jenkins and Brown 2012) have found these types of organizations were often well suited to participate in a range of postdisaster response and relief roles that extended beyond the services they usually provided. For example, one of the shelters Brown has studied routinely helped women navigate multiple crises while simultaneously demonstrating flexibility to overcome limited staff and unanticipated events. The shelter's staff members were particularly adept in helping coordinate social service organizations in the relief effort. Making do with constantly depleting resources, routinely networking within the community to achieve overarching goals, and never knowing what awaited them as they walked in the door helped them improvise when disaster struck (Brown 2009). Routinely working at the "edge of chaos" (Comfort 1999: 14), whether the chaos of people's family lives or the chaos of the waterway, seems to be good training for successfully improvising in disasters.

On their own, the boat operations make for an interesting story, but can we really put our faith in unscripted, decentralized, and emergent disaster management activities? The evidence shows that we can. To do so, however, we—meaning citizens and our officials—have to reconcile the fact that a disaster is an objective and subjective social problem (Wachtendorf, Brown, and Holguin-Veras 2013); that is, there is the physical reality of a particular disaster and something that is perceived, defined, and enacted by those who encounter and make sense of that physicality, sometimes to very different ends.

Rather than figuring how to develop a unified and rational view of disaster, the more appropriate strategy may be to fragment it. An artful fragmentation into discrete parts may provide the needed insight for handling large-scale disasters and catastrophes. That fragmentation cannot be performed by any commanding authority; it must arise from a set of emergent activities.

The 9/11 boat operations are evidence that this sort of response need not be a form of anarchy. Instead, we see a creative distilling of different activities from the universal range of possibilities that constitute disaster. Disasters, and particularly catastrophes, affect entire communities. Consequently, every part of a community can be helpful in a disaster if its involvement is harnessed or allowed to flourish in the right way. Disasters clearly demand certain kinds of activities, such as search and rescue, evacuation, shelter and mass care, coordination of resources, public information, and so on. These are domains ICS attempts to harness. Public agencies do their best to attend to these needs, aided by such nonprofits as the American Red Cross. But it is simply not possible to extend the organizational and technological prowess of these organizations across the *full range of community complexities*—especially in the first confused hours and days following the impact of some major disaster agent. The full panoply of community functioning can hardly be fully coordinated, let alone commanded, in normal times. Consequently, there is no reason at all to think that some predetermined direction can encompass all emergent needs in the disaster context. Instead, having diverse organizations and community institutions tackle disaster needs is the way to go.

Such a stark conclusion does not mean that organizations, or even networks of organizations, cannot function under firm central control. Military organizations deployed to disaster sites answer according to their familiar hierarchical patterns to someone in charge, as do police and fire organizations. But such organizations typically grasp only a segment of the overall disaster, the part that they define as relevant or most immediate to their missions. Search and rescue is vital, for example, but it is only one disaster mission. What those organizations do most successfully is *manage themselves*, a different task from attending to somewhat altered community functions. A system designed to effectively deploy firefighting resources across multiple jurisdictions is not necessarily suitable for delivering coffee, doing laundry, unloading trucks, rounding up golf carts, securing material resources from a host of community organizations, spontaneously boating evacuees across a harbor, or performing any of hundreds of other possible tasks.

Years of experience tell us that organizations are best in disaster situations doing what they normally do and that disaster operations should look as much as possible like normal operations (Dynes 1970, 1994; Quarantelli 1997). When that happens, agencies perform well. The U.S. Coast Guard was universally celebrated following its performance in responding to Hurricane Katrina. What its officers were doing was familiar to them: they rescue people, regardless of whether it is from a boat or from a rooftop. Any organization can perform one or two functions well. But after 9/11, we counted forty-two separate kinds of disaster management functions—let alone the multitude of separate tasks under each—performed in New York City. Nor does this number include the city's usual operations, carried on everywhere north of Canal Street. Many of those functions crystallized in the first few days and carried on for months afterward. Disasters demand organizations that can do everything, but in reality they each cannot do everything.

Organization scholars have defined three kinds of organizational interaction, or *interdependence* (Daft 2004). *Pooled interdependence* has individuals working alongside one another, in parallel, without requiring anything from one another. Think of volunteers picking up trash in a park: they all work independently and need little direction or support, other than direction on where to work, because they are doing something they know how to do. *Sequential interdependence* refers to situations where one individual or group needs something from another before the individual or group can do its work, while *reciprocal interdependence*—the most challenging—means a situation where individuals trade inputs back and forth to function. A disaster management philosophy that looks at disaster as a discrete entity, a single phenomenon, and that tries to manage it as such through a top-down system is, in effect, trying to create sequential and reciprocal interdependences. Every requirement for information or permission creates another interdependence. Such an approach may work in a unified situation, but disasters are milieus of people and places, continuously redefined. Individual functions can be managed, but the *disaster* cannot.

The boat operations might be thought of as clusters of pooled interdependencies that operated with minimal sequential and reciprocal interdependence. Kim Newton did not need anything from Mike Day, who did not need anything from Gerard Rokosz. The actual movement of vessels was self-regulating, as it always was. The operation along Lower Manhattan's waterfront was more complex, because it involved a number of different people performing different functions. Peterson sorted passengers, Day and the pilots coordinated boat traffic (although many of them just coordinated themselves), John Krevey supervised operations at Pier 63, and so on. The bus operators did their own thing at each location where evacuees disembarked. The *John J. Harvey*, resurrecting its identity as a fireboat, interfaced with fire personnel. And meanwhile, an entirely separate disaster response operation was taking place at Ground Zero, where the rest of the world's attention was focused for those first few days.

Louise Comfort (1999); Karl Weick, Kathleen Sutcliffe, and David Obstfeld (1999); John Harrald (2006); and others have provided some clues on how to achieve the seemingly contradictory goal of simultaneously coordinated and flexible action across a large number of entities. A principal requirement is some sort of sharing: of vision, goals, sense of risk, and purpose. Our findings in the World Trade Center disaster response support these broad ideas.

Working in a disaster requires some artistry. Getting a community working again after a disaster is an artistic rather than mechanistic process. As with any art, one employs principles and techniques to create the final work, but handling a disaster is essentially a combination of creative acts by everyone affected. It is an uncomfortable thought, because what we most want in disaster is certainty, including the assurance that people who know what they are doing are on the job. Yet Weick, Sutcliffe, and Obstfeld (1999: 103) argue that an important characteristic of reliable performance in organizations is "enacting moments of organized anarchy":

> They loosen the designation of who is the "important" decision maker in order to allow decision making to migrate along with problems. . . . When problems and decision rights are both allowed to migrate, this increases the likelihood that new capabilities will be matched with new problems.

One boat captain we talked with described the events as "a lot of really good professional mariners doing exactly what they knew how to do—and improvising if we had to improvise." What was it about these mariners that helped them improvise? The mariners themselves saw flexibility as part of what they do day in and day out. One harbor pilot we spoke with put it this way:

> Because that's what we do, you know . . . people who go to sea. . . . You might be the captain of a fishing boat. Well, if the engine breaks, you better be an engineer. . . . I mean, I don't check with somebody to see; I have to make that decision. It's a decision-making process, and people who work in the maritime industry, I believe, are very adaptive at making snap judgments and decisions.

He recalled a conversation with a police captain:

> He told me that we were like Hogan's Heroes, because we would think of some harebrained idea and then . . . we would make it happen. And they couldn't do that. [They had] somebody to answer to. . . . We're very independent. We're not bogged down by hierarchies, usually.

Interestingly, some of the same people who critiqued others for checking in with their supervisors and trying to keep people in their organizational boxes sometimes called for just that. One person told us:

> There was a need for a central figure, a clearinghouse, a group of people who were not bogged down with, "Well, we don't have a contract with that guy" or "Do we have a contract with that guy?" You know, they needed . . . somebody who again would just, like, get the job done, you know, and not worry about whether they'd have a job the next day if [they] made that decision.

The desired role for the leader, it seems, is not to lead in the sense of giving orders or directions but to create or facilitate an environment where people can make sensible decisions. According to one person, many people were not necessarily in charge but rather "go-to people." And that role emerged not necessarily by organizational affiliation but seemingly based on what tasks they were doing, when they started doing them, and how their decisions seemed to fit with what others saw as needing to be done. They just seemed to be there when new situations came up.

The needs of a suddenly devastated city defy organizations' capacities to solely rely on preexisting plans. These needs resist the highly scripted responses in which people pretend that they can solve problems through an omniscient and all-powerful leader. Who, sitting in a command center somewhere in New York, would know that an antique fireboat had the vast pumping capacity needed to substitute for the broken infrastructure at Ground Zero—or, moreover, know that such a boat was actually underway with its faithful crew of volunteers? Who would know that the artful wrought-iron railings along the waterfront were doing a good job of keeping people away from the water and from the boats lined up to rescue them—precisely the opposite of what was needed at that time? An effort to actually control activities would have been counterproductive.

Improvisational actors know that "either/or thinking" blocks creativity. Organizations that can be commanded ought to be commanded. Functions that cannot be commanded ought not to be, but they also should not be derailed. Emergent organizations with their own structural forms can function easily alongside more hierarchical or regimented systems. And when it comes right down to it, the *structure* that is in place is less important than the people who energize the system with their interpretations and decisions—with their sensemaking.

Hierarchical or command-and-control models are for circumstances requiring genuine coordination, especially at a tactical level—manipulating heavy equipment around Ground Zero, for example, or situations where people or resources need to be tracked to avoid interfering or endangering one another's work. Overwhelmingly, most disaster response functions do not require that level of coordination. Fortunately, in most cases, the spontaneous volunteers and assorted helpers who appear at disaster sites will not be trying to get the top-down view that officials crave. They will happily occupy their roles in different areas—Kim Newton at Highlands, Gerard Rokosz at Weehawken, or Jeff Wollman chugging through Lower Manhattan in a sputtering truck. The disaster response power of those activities is that people took over pieces of the disaster as their own milieu.

Sucking all those functions into ICS would have been pointless and probably impossible. Some participants who were familiar with ICS mentioned it when we talked with them, but even for them, it seems at most to have been a kind of mental model or framework, since the system itself was not used in the boat evacuation. They did not establish formal ICS sections. One pilot with extensive ICS training said, "We didn't really have time for that." Another noted that most of the people he was in contact with did not know anything about it. Another participant commented, "Nobody got assigned anything, especially on this site. . . . We worked [it] out." As one person put it:

> There was no ICS structure. . . . ICS, we're very familiar with the training. That wasn't the case. [Rather it] was self-actualization. It's the highest form of emergency management, because people would see something that needed to be done and they would do it.

Concerns that emergent and loosely controlled activities will cause interference or draw resources more urgently needed elsewhere do not seem to have panned out in the boat operations. Traditionally, emergency managers have been particularly worried about the managerial challenges of supervising volunteers. An emergency manager once stated:

> The more volunteers, the more fractured the structure, the more other departments are involved, the more difficult the management problem. You need people you know and trust with whom you have instant communications. The next layer of people, whether the Army or the Kiwanis Club, creates as many problems as it solves. (Kartez and Kelley 1988: 140)

But volunteers create problems only if we see their work as something that needs to be integrated into a unified system, where people need to be supervised. If we were to offer guidance to officials, it would be to allow as much pooled interdependence as possible and to avoid creating complications or occasions for conflict that occur with sequential or reciprocal interdependence between components (Daft 2004). Over time, for example, Day turned away smaller boat operators who had little passenger or cargo capacity. Peterson made forays to Ground Zero to meet with officials, while independent operations sprang up at various locations. None of these people had much contact with one another.

The lack of a command structure was frustrating to some. As one person put it, "People just wanted to be put to work. You know, a lot of people, they didn't want to think. Don't make them make a decision . . . but you need people to tell them what to do." He elaborated, "[I] wanted to find some structure, and I kept looking for the structure, and I kept finding no structure, and the only structure that was there was structure that people made." But many times that emergent structure worked. As one ferry operator stated, "It was just like a lot of good people just getting together, leaning into it, and like forming our own little ring structure." Another mariner stated, "If somebody saw a need for something, they'd just, you know, look around, and if nobody else was doing it, they'd hop in and do it."

Our findings do not yield themselves to a checklist of procedures for managing a disaster. Instead, they suggest the necessity of a certain degree of flexibility in disaster management philosophy. We know there will always be emergence in large events. We suggest that official organizations need to gain an appreciation for improvisation and a tolerance for decentralization and that each quality should be part of the professional training and professional virtues of emergency officials.

Role of Planning

The boat operations prove that when there is familiarity and acquaintance among a group of potential convergers, plans and prescripted systems are not necessarily needed. Indeed, it is not entirely clear whether it would have been possible to plan for something like the evacuation of Lower Manhattan. The boat operators themselves might have objected to such ideas as carrying dozens of people on a tugboat or loading passengers without a gangplank had they been asked to participate in a planning process beforehand. Nevertheless, we do not want this book to be read as an argument against planning. Planning and improvising are not exclusive concepts. Gary Kreps (1991: 4), a sociologist who is an expert on organizational improvisation, has suggested that planning is organizing in advance, while improvising is organizing in the present.

One of the mariners we talked with put it this way:

> A plan to me is a range of options. A plan isn't what I'm going to do. How I view planning is [that] it's a realm of options and I'm going to choose what works best for me in a particular time. [Some people] believe plans are to be written in the calm of day and it's going to have all the answers. . . . I think it gives you things to think about. . . . But to say, "First we're going to do this, and then we're going to do this; then we're going to go here. . . ." I did one year of planning, and I remember typing until my eyes were bloodshot, all this stuff and [planning this and that], and then something happened, and [the plan] was just totally disregarded. That was my first few experiences with planning, and I always thought plans should be different. They should give you an option.

We should think about such matters as evacuations and staging areas, but we should also recognize that something might happen to cause those plans to fall apart. In those cases, we need to compromise on our solutions to the problem. As one participant said, "Why not train ourselves to think along with a number of alternatives and scenarios?"

But what of situations where there is a plan and people do not follow it, and the result is a mess? Shouldn't they have followed the plan? We certainly cannot say that every deviation from a plan is a good decision, that every emergent or independent action is necessarily good. Some plans obviously work. If a plan is working, fine. But if there is no plan, or if the plan is not working, spontaneous action is needed to fill the gaps. Plans are always deficient in disaster—and even in ordinary times. Plenty of studies have shown that the way work really gets done is different from how bosses think the work gets done, and studies have shown that the skills workers think are important differ from those that managers think are important (Brown and Duguid 1994; Darrah 1994). If even familiar systems have such a strong degree of approximation, what of disaster: a milieu of people and places that is a genuinely unfamiliar? Familiarity, acquaintance, environmental knowledge, a sense of identity and purpose leading to partnerships, imagination and creativity, underlying sensemaking and wisdom: these are the real needs in disaster. No plan is a substitute for these fundamentals.

Another mariner said, "Dealing with Mother Nature, you always get a curveball thrown at you, so you get used to thinking on your feet rather than just sitting here and plotting out what we are going to do for the day." As prepared as the *Harvey* was, the challenge its crew faced was that the *Harvey* was set up to shoot water out of its deck pipes, not into hoses, and the valves were not set up to work in that way. Tim Ivory was ready to improvise, however, by shoving Coke bottles and softballs into the guns they were not using, diverting water through the pipes the way they needed it.

One of the reasons that officials pin their hopes on centralized management structures (and plans) is their belief that such structures can limit the possibility for conflicting demands for resources—basically, that a more nimble or more powerful agency will rob another, even if the latter's needs are greater. They may also fear that resources will be duplicated in one place while shorted in another, as happened during the 1970 California wildfires (Irwin 1989). The concern is legitimate. And, in fact, while observing decisions in New York City's EOC, we saw officials who were more vocal, more persuasive, or more skilled get more attention from higher-ups than those who were less so. We think that this kind of politicking introduces its own randomness into the decision process, because it produces outcomes that may be disconnected from ground-level conditions. No system or plan can be effective if the people operating it are not alert to the formal power structures and informal eddies of influence that exist in all organizations.

The process of organizing resources may ultimately be as central to managing a disaster as establishing structures; this, too, can be difficult to plan for. There were not enough government boats available to handle the approximately five hundred thousand people who needed to get out of Manhattan on 9/11. Instead, the evacuation depended on the resources of the commercial maritime community and the commercial bus community. Moreover—in keeping with our concept of disaster as a product of reality and of imagination—it is important to note that resources and skills do not exist as things in and of themselves. Nick Middleton (1999: 17), a geographer, has argued, "Since resources are simply a cultural appraisal of the material world, individual aspects of the environment can vary at different times and in different places between being resources and negative resources."

Skills, too, are a kind of resource that is simultaneously socially constructed and real. Jamie Peck (1996: 135) argues that "skill should not be seen simply as a resource that is rewarded in accordance with the precepts of human capital theory, but as an ideological construct reflecting the distribution of power in the labor market." In other words, skills are not just *there*; they are created and understood relative to market needs. In a disaster, the market for potential skills is vast. The participation of civil society in disaster response efforts broadens the range of potential skills, in effect defining needs through popular participation. Bicycle couriers made a market for themselves, as did chiropractors, food vendors, bartenders handing out chips, and others, and massage therapists enormously increased in visibility because of the assistance that they gave workers on the pile at Ground Zero. Far from creating resource conflicts, emergent responses grounded in civil society's interpretations of disaster expand the pool of people and resources.

We are often asked what the boat evacuation really means. Was it just a one-off event, a product of the unique capacities of New Yorkers, or of mariners in particular? Certainly, there are elements that are specific to New York and specific to the maritime environment. But, as Russell Dynes (2003) writes, people in different places have rebounded from terrible assaults and destruction. Whether we can predict a similarly adaptive response in every place struck by disaster is an open question, but on the odds, based on historical experience, people will try to help themselves and others, at least in the early days, before the conflicts of the recovery process ensue. They will be determined and resourceful in defining their mission. A more useful question, therefore, is what makes the move toward helping productive and actually helpful.

Identity is important for sparking the initial sense among some potential volunteers that they can potentially help with an ongoing problem. A lot of people can conceive of the same idea at the same time, and that certainly happened in New York, as members of the civilian maritime community and the Coast Guard realized that boats offered the best way to clear people from the waterfront. Together, their emergent activity suggested possibilities or obligations for others. Having the ability to work unsupervised, to break a problem down into bite-size pieces, to make useful alliances, and to avoid burdensome connections seem to be key to successful participation. Peterson framed his own problem and limited his responsibilities to evacuees around Lower Manhattan, trusting that someone else would solve the problem of getting the passengers to their ultimate destinations. The "system" that emerged was tacit, indexed through people's identities, skills, and locations, and thus was truly modular.

These sorts of activities cannot be commanded into existence; they stem from the same norms that create our communities. What we are looking at, then, is less a prescriptive model of disaster response than a concept of how disaster activities can unfold. It is a milieu of multiple problem-solving systems: the highly organized and hierarchical models of designated response agencies, emergent modules of activities in social spaces those agencies cannot reach, and boundary-spanning activities between these domains to share knowledge and resources as needed. The concept is highly improvisational, blending planned activities and emergent inspiration.

The existing research literature suggests that planning and improvisation have an inverse relationship: the more *good planning* there is, the less need for improvisation. In some senses, this is true. The more people plan in the direction of foreseeable outcomes and needs, the less they will have to improvise in situations with challenges that can be defined in advance. But while disasters may bring about *predictable needs*—search and rescue, evacuation, shelter—they also tend to bring about *unpredictable shifts in the preconditions* for meeting those needs. Understanding those preconditions is perhaps more important to the planning process than identifying what are, by now, well-established emergency management functions. It is, however, also the most difficult, because it requires a detailed mapping of community resources and capabilities, features that often change—sometimes quite substantially—over time.

There is a strong normative dimension to the apparent conflict between planning and improvisation. Improvisation automatically suggests a failure to plan, which in turn suggests a failure of competence. But our communities invariably follow unstable developmental and functional paths, and this process is not likely to be any different in disaster. Thus, instead of automatically seeing improvisation as an admission of failure, people should see improvisation as an adaptive response to new conditions in the natural, technical, and social environments. Is it possible to plan to improvise?

Our observations of the World Trade Center disaster response suggest that public officials need to take a different look at planning than has previously been evident. Part of the examination of every large disaster has generally included a call for "better planning." These calls suggest that all the dimensions of the disaster could have been identified in advance. Certainly a broad range of *needs* can be identified. However, identifying such needs is different from identifying the wherewithal to accomplish the tasks required to fulfill those needs. Planning is essential, but it is only a "hypothesis" or a supposition about the form that a combination of future organizational structures and resources is likely to take (Michael McGuire, pers. comm.).

We believe that the evidence provided by large-scale events suggests that training should offer more in the way of fostering capacities for creativity and improvisation. A disaster, by definition, requires improvisation (Tierney 2002); as the magnitude of the disaster increases, greater levels of innovative thinking are needed. Plans should provide guidelines for actions and desired outcomes, and they should outline relationships and zones of responsibility. They are less effective at projecting specific response actions into an imagined future of greater and greater levels of disaster.

Consider the 2010 Deepwater Horizon oil spill, an operation that pushed drilling technology to the maximum extent and where, in the event of trouble, emergency solutions were themselves experimental. As we subsequently learned, in those situations, at those depths, safety had to be front-loaded into the system. Everything had to work right upon installation, because retrofitting a solution was nearly impossible. Lee Clarke (1999: 16) has argued that plans for catastrophic oil clean-up are "fantasy documents," because the technical capacity, organization prowess, and historical experience do not exist. With scientifically exotic events, such as Deepwater Horizon, the creativity may be technical. In addition to technical creativity (e.g., developing systems to support the underground walls of the Trade Center complex as debris was removed), September 11 required massive levels of organizational creativity and initiative.

How is it possible to train to be creative? Creativity is a large and complex subject; researchers differ on the elements of individual and organizational creativity. Most researchers agree, however, that people actually tend toward creativity (Comfort 1999) and try to act creatively when they can. One useful first step in rethinking disaster response may be to eliminate the barriers to creativity so that people can make use of their innate creative impulses. In the specific case of disaster training, training for creativity would have much less to do with learning a plan and following its steps—although deep familiarity is important—and more to do with practicing interacting with the various organizations that would be involved in response. Such training would be more action-oriented and would emphasize increased problem-solving capacity rather than procedural rectitude. Without question, such training would be difficult, and probably more expensive. Nevertheless, organizations need practice in working together, in searching for new, unusual, or unexpected sources of materials and assistance, and, perhaps most importantly, they need practice in looking imaginatively at an evolving situation.

The training process and the planning process are intimately intertwined. Planning should be conceived of as training, since it is in the planning process that future participants in disaster operations will become familiar with one another's resources, capabilities, and even very existence. Most disaster scholars agree that the planning process is more important than the actual written plan, and it is now axiomatic that a "plan" is meaningless if the people who will be implementing it have had no role in its production. Planning and training should be simultaneous and should more correctly be called an action-learning process, where a plan is the outcome of organizations working together in problem solving (Dynes 2003). By increasing the number of participants, it is possible to spread potential problems across a wider array of perspectives and maximize the chances for finding solutions, a principle that researchers note is important in "high-reliability organizations" (Weick, Sutcliffe, and Obstfeld 1999: 81). And even then, the process should appreciate that some of those who will ultimately have critical roles to play in the disaster response milieu will not have been at the table.

And what of regular people? We could, of course, take the view that disaster victims are a helpless lot; that people in a community affected by disaster are victims; that victims are vulnerable and, therefore, are helpless. We could take that view, but if we did, we would be wrong. People can simultaneously need assistance while at the same time demonstrate strong capacities to provide assistance to others. Although it may be easier to mentally map victims and helpers into dichotomous groups, our research demonstrates that community disasters provide space for its members to engage as agents of response.

In our day-to-day activities, we often rely on the background of community functioning. We do not give much thought to the bus or ferry showing up on schedule, the traffic lights changing color, or the person who will unlock the doors to the school in the morning. These operations become visible only when they cease to function as expected. But there are those in our community who know these functions well because they attend to them every day. These community members have the sense to determine what needs to be done and have the skills to do it during a disaster, even when the rest of us have yet to imagine those needs. Be it identifying the need to evacuate people by boat, or hand out chips with a comforting smile along a pier, or check on a neighbor we know might require some assistance, we all have identities and skills that can be of use when our communities are affected by disaster.

We can be mindful in our initial muddling. We can get out of the way when others can better handle what needs to be done, as some smaller vessels did or as the Coast Guard did when other boat operators were doing well on their own. We can also look for the need that perhaps no one else yet sees. Others with a similar identity or set of experiences may be doing the same job in other places. If we focus too much on the extraordinary, disaster can become a monumental challenge that appears impossible to overcome. But in New York, the extraordinary response came about from people turning ordinary activity into something they could do and manage. By fostering the ordinary in the extraordinary, we will work to solve at least some of the challenges of disaster response.

Acknowledgments

This book began on 9/11. The research on which it is based has evolved from an initial chapter in a volume published by the Natural Hazards Center at the University of Colorado, and our findings have been presented in multiple scholarly, practice, and policy venues.

Funds for portions of this research were provided by the Multidisciplinary Center for Earthquake Engineering Research (MCEER) New Technologies in Emergency Management No. 00-10-81 and Measures of Resilience No. 99-32-01 through National Science Foundation award No. 9701471, the Public Entity Risk Institute No. 2001-70 (Kathleen Tierney, Principal Investigator), National Science Foundation No. 0603561 and No. 0510188 (James Kendra and Tricia Wachtendorf, Principal Investigators), and the University of Delaware Research Foundation (Tricia Wachtendorf, Principal Investigator). We are grateful to the South Street Seaport Museum (Jeffrey Remling, Collections Director) for access to interviews with participants in the waterborne operations. Funding to the museum for these interviews was provided by the National Endowment for the Humanities, and the interviews were conducted by David Tarnow.

We thank the University of North Texas and the University of Delaware for sabbatical leaves during which we worked on this book, and our departments at these universities for their encouragement and support of our research: the University of North Texas Department of Public Administration and the University of Delaware School of Public Policy and Administration and Department of Sociology and Criminal Justice. We are also grateful to the CRISMART (Swedish National Defense College) for space provided during one of these sabbaticals.

We were not alone in our research efforts. Jasmin Ruback was a valued colleague throughout the project. Graduate students John Barnshaw, Brandi Lea, and Lynn Letukas conducted some of the interviews. They, along with Alicia Baddorf, Austin Barlow, Lauren Barsky, Bethany Brown, Mary Chaffee, Jeff Engle, Brandy Gilbert, Scott Golden, Joshua Kelly, Mary McColley, Margaret McNeil, Cynthia Rivas, Lauren Ross, Brittany Scott, Gabriela Wasileski, and Caroline Williams, helped with the collection and compilation of data for this project. Their contributions were valuable and are much appreciated.

We would be remiss if we failed to offer our gratitude to several individuals. First, we thank Kathleen Tierney, who has been a good friend, colleague, and mentor for many years and without whose help we would not have been able to embark on this research. We are grateful to Lee Clarke, who gave his time generously to read various drafts of papers, which, in turn, informed this research. And we are similarly grateful to Ken Mitchell, who directed a dissertation that formed some of the background knowledge on maritime management. Disaster Research Center (DRC) founding directors Enrico L. Quarantelli and Russell Dynes were ever encouraging as this project evolved, and the library and collection management skills of DRC resource coordinator Pat Young greatly facilitated the completion of our work. Victoria Becker, DRC's business manager, contributed valuable financial management skills. Richard Rotanz was an early supporter of our work in New York City and is a continuing advocate for the value of research to the practice of disaster management. We hope that the findings presented here, embedded in the compelling story that the waterborne evacuation provided, live up to those standards. We also thank Audra Wolfe, a skilled science editor, who provided many recommendations and suggestions for improving the first draft of this book (and who bears no responsibility for whether or how well we followed her advice). We are very grateful to Heather Wilcox, a vigilant editor and proofreader, who caught typos and inconsistencies of every sort. Joy Dean Lee artfully indexed the content. We appreciate the support and patience of Micah Kleit, our editor at Temple University Press. The world of disaster research does not function on a time line, and we were often diverted by new disasters and emergencies. Micah granted generous extensions as deadlines passed.

Jim offers many thanks to April for her ongoing love and encouragement and for putting up with his absences to work on this book, especially in the last weeks as the actual, serious, genuine deadline for submission approached.

Tricia thanks David for all things helpful, distracting, practical, and impractical. And she thanks Shoshi and Ari for being the loving and crazy wack-a-hoos they are.

Finally, we owe a great debt of gratitude to all the maritime workers who shared their recollections for this project. On their boats, in restaurants, and on the phone, they gave their time to us, and without them we could not have written this book. We hope that we have succeeded in reflecting their stories with sensitivity and understanding.

References

Aguirre, Benigno E. 1994. *Planning, Warning, Evacuation and Search and Rescue: A Review of the Social Science Research Literature*. College Station, TX: Hazard Reduction and Recovery Center, College of Architecture, Texas A&M University.

Aguirre, Benigno E., Dennis Wenger, and Gabriela Vigo. 1998. "A Test of Emergent Norm Theory of Collective Behavior." *Sociological Forum* 13 (2): 301–320.

Aichele, Richard O. 2002. "A Shining Light in Our Darkest Hour." *Professional Mariner* 61. Available at http://www.fireboat.org/press/prof_mariner_jan02_1.asp.

Alexander, David A. 1993. *Natural Disasters*. New York: Chapman and Hall.

———. 2012. "What Can We Do about Earthquakes? Towards a Systematic Approach to Seismic Risk Mitigation." Global Risk Forum (GRF), Davos, Switzerland. Available at http://www.nzsee.org.nz/db/2012/Paper001.pdf.

Alway, Joan, Linda Liska Belgrave, and Kenneth J. Smith. 1998. "Back to Normal: Gender and Disaster." *Symbolic Interaction* 21 (2): 175–195.

Auf der Heide, Erik. 1989. *Disaster Response: Principles of Preparation and Coordination*. St. Louis: Mosby.

Barton, Allen H. 1969. *Communities in Disaster: A Sociological Analysis of Collective Stress Situations*. Garden City, NY: Doubleday.

Beunza, Daniel, and David Stark. 2003. "A Desk on the 20th Floor: Survival and Sense-Making in a Trading Room." Working Paper Series, Center on Organizational Innovation, Columbia University, New York. Available at http://www.coi.columbia.edu/pdf/beunza_stark_d20.pdf.

Bok, Sissela. 1978. *Lying: Moral Choice in Public and Private Life*. New York: Pantheon.

Brown, Bethany. 2009. *Organizational Response and Recovery of Domestic Violence Shelters in the Aftermath of Disaster.* Newark: University of Delaware.

———. 2012. "Battered Women's Shelters in New Orleans: Recovery and Transformation." In *The Women of Katrina: How Gender, Race, and Class Matter in an American Disaster*, edited by Emmanuel David and Elaine P. Enarson, 179–189. Nashville: Vanderbilt University Press.

Brown, Bethany L., Pamela Jenkins, and Tricia Wachtendorf. 2010. "Shelter in the Storm: A Battered Women's Shelter and Catastrophe." *International Journal of Mass Emergencies and Disasters* 28 (2): 226–245.

Brown, John Seely, and Paul Duguid. 1994. "Borderline Issues: Social and Material Aspects of Design." *Human Computer Interaction* 9 (1): 3–36.

Buck, Dick A., Joseph E. Trainor, and Benigno E. Aguirre. 2006. "A Critical Evaluation of the Incident Command System and NIMS." *Journal of Homeland Security and Emergency Management* 3 (3): 1–29.

Chong, Jia-Rui. 2009. "Study Finds Troubling Pattern of Southern California Quakes." *Los Angeles Times*, January 24. Available at http://articles.latimes.com/2009/jan/24/local/me-fault-quakes24.

Clarke, Lee. 1999. *Mission Improbable: Using Fantasy Documents to Tame Disaster.* Chicago: University of Chicago Press.

CNN. 2005a. "Military Due to Move In to New Orleans." CNN.com, September 2. Available at http://www.cnn.com/2005/WEATHER/09/02/katrina.impact/.

———. 2005b. "CNN Presents: CNN Security Watch: Lessons of Hurricane Katrina." CNN.com, September 10. Available at http://transcripts.cnn.com/TRANSCRIPTS/0509/10/cp.01.html.

———. 2005c. "Mayor to Feds: 'Get Off Your Asses.' Transcript of Radio Interview with New Orleans' Nagin." CNN.com, September 2. Available at http://www.cnn.com/2005/US/09/02/nagin.transcript/.

Comfort, Louise K. 1999. *Shared Risk: Complex Systems in Seismic Response.* Pittsburgh, PA: Pergamon.

Daft, Richard L. 2004. *Organization Theory and Design.* 8th ed. Mason, OH: South-Western.

Darrah, Charles. 1994. "Skill Requirements at Work: Rhetoric versus Reality." *Work and Occupations* 21 (1): 64–84.

Denhardt, Robert B., Janet V. Denhardt, and Maria P. Aristigueta. 2009. *Managing Human Behavior in Public and Nonprofit Organizations.* 2nd ed. Thousand Oaks, CA: Sage.

Dombrowsky, Wolf R. 1998. "Again and Again: Is a Disaster What We Call a Disaster?" In *What Is a Disaster: Perspectives on the Question*, edited by Enrico L. Quarantelli, 19–30. London: Routledge.

Drabek, Thomas E. 1996. *Disaster Evacuation Behavior: Tourists and Other Transients.* Boulder: University of Colorado, Institute of Behavioral Science.

Drabek, Thomas E., and David A. McEntire. 2002. "Emergent Phenomena and Multiorganizational Coordination in Disasters: Lessons from the Research Literature." *International Journal of Mass Emergencies and Disasters* 20 (2): 197–224.

"Duty to Provide Assistance at Sea." 2010. 46 U.S. Code § 2304. Available at https://www.gpo.gov/fdsys/granule/USCODE-2009-title46/USCODE-2009-title46-subtitleII-partA-chap23-sec2304.

Dynes, Russell R. 1970. *Organized Behavior in Disaster.* Lexington, MA: Heath Lexington Books.

———. 1994. "Community Emergency Planning: False Assumptions and Inappropriate Analogies." *International Journal of Mass Emergencies and Disasters* 12 (2): 141–158.

———. 2003. "Finding Order in Disorder: Continuities in the 9-11 Response." *International Journal of Mass Emergencies and Disasters* 21 (3): 9–23.

Enarson, Elaine. 2001. "What Women Do: Gendered Labor in the Red River Valley Flood." *Global Environmental Change Part B: Environmental Hazards* 3 (1): 1–18.

Fireboat.org. 2015. Available at http://www.fireboat.org/.

Fischer, Henry W., III. 2008. *Response to Disaster: Fact versus Fiction and Its Perpetuation.* 3rd ed. Lanham, MD: University Press of America.

Fordham, Maureen. H. 1998. "Making Women Visible in Disasters: Problematising the Private Domain." *Disasters* 22 (2): 126–143.

Fritz, Charles E. 1961. "Disasters." In *Contemporary Social Problems: An Introduction to the Sociology of Deviant Behavior and Social Disorganization,* edited by Robert King Merton and Robert A. Nisbet, 651–694. New York: Harcourt, Brace and World.

Fritz, Charles E., and John H. Mathewson. 1957. *Convergence Behavior in Disasters: A Problem in Social Control.* Washington, DC: National Academy of Sciences, National Research Council.

Grzyb, Gerard J. 1990. "Deskilling, Decollectivization, and Diesels: Toward a New Focus in the Study of Changing Skills." *Journal of Contemporary Ethnography* 19 (2): 163–187.

Harrald, John R. 2006. "Agility and Discipline: Critical Success Factors for Disaster Response." *Annals of the American Academy of Political and Social Science* 604 (1): 256–272.

Hayler, William B., John M. Keever, Paul M. Seiler, and Felix M. Cornell, eds. 1980. *American Merchant Seaman's Manual: For Seamen by Seamen.* 6th ed. Centreville, MD: Cornell Maritime Press.

"Hero of Utoya Island: German Holidaymaker Saved 20 Youngsters from Massacre Using Rented Boat." 2011. *Daily Mail,* July 25. Available at http://www.dailymail.co.uk/news/article-2018479/Norway-massacre-Marcel-Gleffe-saved-20-teenagers-Anders-Behring-Breivik.html#ixzz1pZwT0fGP.

Holland America Line. 2015. *Onboard Safety.* Available at http://www.hollandamerica.com/virtual-tours-videos/Main.action?cat=ships&subcat=am&type=video&id=4&title=Holland%20America%20Cruise%20Line%20Video:%20Onboard%20Safety&desc=.

Hughes, Everett C. 1945. "Dilemmas and Contradictions of Status." *American Journal of Sociology* 50 (5): 353–359.

Hutchins, Edwin. 1994. "In Search of a Unit of Analysis for Technology Use." *Human Computer Interaction* 9 (1): 78–81.

Irwin, Robert L. 1989. "The Incident Command System (ICS)." In *Disaster Response: Principles of Preparation and Coordination*, 133–163. St. Louis: Mosby.

Jenkins, Pamela, and Bethany Brown. 2012. "Rebuilding and Reframing: Nonprofit Organizations Respond to Hurricane Katrina." In *Crime and Criminal Justice in Disaster*, edited by Dee Wood Harper and Kelly Frailing, 339–358. 2nd ed. Durham, NC: Carolina Academic Press.

Kartez, Jack D., and William J. Kelley. 1988. "Research-Based Disaster Planning: Conditions for Implementation." *Managing Disaster: Strategies and Policy Perspectives*, edited by Louise K. Comfort, 126–146. Durham, NC: Duke University Press.

Kelly, John, and David Stark. 2002. "Crisis, Recovery, Innovation: Responsive Organization after September 11." *Environment and Planning A* 34:1523–1533.

Kendra, James M. 2000. "Looking Out the Window: Risk, Work, and Technological Change in US Merchant Shipping." PhD diss., Rutgers University, New Brunswick, NJ.

———. 2007. "The Reconstitution of Risk Objects." *Journal of Risk Research* 10 (1): 29–48.

Kendra, James M., and Tricia Wachtendorf. 2003. "Reconsidering Convergence and Converger Legitimacy in Response to the World Trade Center Disaster." *Research in Social Problems and Public Policy* 11:97–122.

———. 2007. "Improvisation, Creativity, and the Art of Emergency Management." *Understanding and Responding to Terrorism*, edited by Huseyin Durmaz, Bilal Sevinc, Ahmet Sait Yayla, and Siddik Ekici, 324–335. Washington, DC: IOS Press.

Knowles, Scott Gabriel. 2011. *The Disaster Experts: Mastering Risk in Modern America*. Philadelphia: University of Pennsylvania Press.

Kreps, Gary A. 1991. "Organizing for Emergency Management." In *Emergency Management: Principles and Practice for Local Government*, ed. Thomas E. Drabek and Gerard J. Hoetmer, 30–54. Washington, DC: International City Management Association.

Lagadec, Patrick. 1993. Preventing Chaos in a Crisis: Strategies for Prevention, Control, and Damage Limitation. London: McGraw-Hill.

Lewis, Russell. 2015. "Remembering Apollo 13, NASA's Most Famous 'successful failure.'" NPR.org, April 10. Available at http://www.npr.org/2015/04/10/398824586/remembering-apollo-13-nasas-most-famous-successful-failure.

Lindblom, Charles E. 1959. "The Science of 'Muddling Through.'" *Public Administration Review* 19 (2): 79–88.

Mallak, Larry. 1998. "Resilience in the Healthcare Industry." Paper presented at

7th Annual Industrial Engineering Research Conference, May 9–10, Banff, Alberta, Canada.

Middleton, Nick. 1999. *The Global Casino: An Introduction to Environmental Issues*. 2nd ed. London: Arnold.

Miller, Katherine. 1998. "The Evolution of Professional Identity: The Case of Osteopathic Medicine." *Social Science and Medicine* 47 (11): 1739–1748.

Mitchell, James Kenneth. 1990. "Human Dimensions of Environmental Hazards: Complexity, Disparity, and the Search for Guidance." In *Nothing to Fear: Risks and Hazards in American Society*, edited by Andrew Kirby, 131–175. Tucson: University of Arizona Press.

———, ed. 1996. "Improving Community Responses to Industrial Disasters." In *The Long Road to Recovery: Community Responses to Industrial Disasters*, 10–40. New York: United Nations University Press.

———. 1999. "Hazards and Culture: New Theoretical Perspectives." Paper presented at the 95th Annual Meeting of the Association of American Geographers, Honolulu, Hawaii, March 23–27.

———. 2003. "The Fox and the Hedgehog: Myopia about Homeland Vulnerability in US Policies on Terrorism." *Research in Social Problems and Public Policy* 11:53–72.

———. 2006. "The Primacy of Partnership: Scoping a New National Disaster Recovery Policy." *Annals of the American Academy of Political and Social Science* 604 (1): 228–255.

Mostert, Noël. 1974. *Supership*. New York: Knopf.

Moynihan, Donald P. 2009. "The Network Governance of Crisis Response: Case Studies of Incident Command Systems." *Journal of Public Administration Research and Theory* 19 (4): 895–915.

National Commission on Terrorist Attacks upon the United States. 2004. *The 9/11 Commission Report: Final Report of the National Commission on Terrorist Attacks upon the United States*. New York: Norton.

National Research Council (U.S.), Committee on Advances in Navigation and Piloting. 1994. *Minding the Helm: Marine Navigation and Piloting*. Washington, DC: National Academy Press.

Neal, David M., and Brenda D. Phillips. 1995. "Effective Emergency Management: Reconsidering the Bureaucratic Approach." *Disasters* 19 (4): 327–337.

Ogintz, Eileen. 2012. "What You Need to Know about Kids' Safety on a Cruise." FoxNews.com, February 10. Available at http://www.foxnews.com/travel/2012/02/10/taking-kids-what-need-to-know-about-cruise-safety/.

Paton, Douglas. 2003. "Stress in Disaster Response: A Risk Management Approach." *Disaster Prevention and Management: An International Journal* 12 (3): 203–209.

PBS Home Video. 2002. *America Rebuilds: A Year at Ground Zero*. Directed by Seth Kramer and Daniel A. Miller. Great Projects Film Company, Shadowbox Films, Inc., Trigger Street Productions.

Peacock, Walter Gillis, and Kathleen Ragsdale. 1997. "Social Systems, Ecological

Networks, and Disasters: Toward a Socio-political Ecology of Disasters." In *Hurricane Andrew: Ethnicity, Gender, and the Sociology of Disasters*, edited by Walter Gillis Peacock, Betty Hearn Morrow, and Hugh Gladwin, 20–35. London: Routledge.

Peck, Jamie. 1996. *Work-Place: The Social Regulation of Labor Markets*. New York: Guilford Press.

Peek, Lori A. 2011. *Behind the Backlash: Muslim Americans after 9/11*. Philadelphia: Temple University Press.

Perrow, Charles. 1977. "The Bureaucratic Paradox: The Efficient Organization Centralizes in Order to Decentralize." *Organizational Dynamics* 5 (4): 3–14.

———. 1984. *Normal Accidents: Living with High-Risk Technologies*. New York: Basic Books.

Perry, Ronald W. 2006. "What Is a Disaster?" In *Handbook of Disaster Research*, edited by Havidán Rodríguez, Enrico L. Quarantelli, and Russell R. Dynes, 1–15. New York: Springer.

Quarantelli, Enrico L. 1997. "Ten Criteria for Evaluating the Management of Community Disasters." *Disasters* 21 (1): 39–56.

———. 2000. "Emergencies, Disasters, and Catastrophes Are Different Phenomena." Preliminary Paper #304, 1–5. Newark: University of Delaware, Disaster Research Center.

———. 2006. "Catastrophes Are Different from Disasters: Some Implications for Crisis Planning and Managing Drawn from Katrina." In *Understanding Katrina: Perspectives from the Social Sciences*. Available at http://understandingkatrina.ssrc.org/Quarantelli/.

Rochlin, Gene I. 1989. "Informal Organizational Networking as a Crisis Avoidance Strategy: U.S. Naval Flight Operations as a Case Study." *Industrial Crisis Quarterly* 3 (2): 159–176.

Rodríguez, Havidán, Joseph Trainor, and Enrico L. Quarantelli. 2006. "Rising to the Challenges of a Catastrophe: The Emergent and Prosocial Behavior Following Hurricane Katrina." *Annals of the American Academy of Political and Social Science* 604 (1): 82–101.

Snyder, John. 2001. "Ferries to the Rescue after World Trade Center Terror Attack." *Marine Log*, October. Available at http://www.marinelog.com/DOCS/PRINT/mmiocfer1.html.

Stallings, Robert A., and Enrico L. Quarantelli. 1985. "Emergent Citizen Groups and Emergency Management." *Public Administration Review* 45:93–100.

Sylves, Richard T. 2007. "U.S. Disaster Policy and Management in an Era of Homeland Security." In *Disciplines, Disasters and Emergency Management the Convergence and Divergence of Concepts: Issues and Trends from the Research Literature*, edited by David A. McEntire, 142–160. Springfield, IL: Charles C. Thomas.

Tennyson, Alfred, Lord. (1854) 1958. "The Charge of the Light Brigade." In *Poems of Tennyson*, edited by Jerome Buckley, 274–276. Cambridge, MA: Riverside Press.

Tierney, Kathleen J. 2002. "Lessons Learned from Research on Group and Organizational Responses to Disasters." Paper presented at Countering Terrorism: Lessons Learned from Natural and Technological Disasters, National Academy of Sciences, Washington, DC, February 28–March 1.

———. 2003. "Disaster Beliefs and Institutional Interests: Recycling Disaster Myths in the Aftermath of 9-11." In *Terrorism and Disaster: New Threats, New Ideas: Research in Social Problems and Public Policy*, edited by Lee Clarke, 33–51. Bingley, UK: Elsevier.

Tierney, Kathleen, Christine Bevc, and Erica Kuligowski. 2006. "Metaphors Matter: Disaster Myths, Media Frames, and Their Consequences in Hurricane Katrina." *Annals of the American Academy of Political and Social Science* 604 (1): 57–81.

Tierney, Kathleen J., Michael K. Lindell, and Ronald W. Perry. 2001. *Facing the Unexpected: Disaster Preparedness and Response in the United States.* Washington, DC: Joseph Henry Press.

Trainor, Joseph E., and Lauren E. Barsky. 2011. "Reporting for Duty? A Synthesis of Research on Role Conflict, Strain, and Abandonment among Emergency Responders during Disasters and Catastrophes." University of Delaware, Disaster Research Center Miscellaneous Report #71.

Transportation Safety Board (TSB). 1998. "Marine Occurrence Report. Near-Collision between the Cruise Ship 'STATENDAM' and the Tug/Barge Unit 'BELLEISLE SOUND'/'RADIUM 622' Discovery Passage, British Columbia 11 August 1996." Report Number M96W0187. Available at http://www.tsb.gc.ca/ENG/reports/marine/1996/em96w0187.html.

———. 1999. "Marine Occurrence Report: Grounding: The Bulk Carrier 'RAVEN ARROW'/Johnstone Strait, British Columbia/24 September 1997." Report Number M97W0197. Available at http://www.tsb.gc.ca/eng/rapports-reports/marine/1997/m97w0197/m97w0197.pdf.

Turner, Ralph H., and Lewis M. Killian. 1987. *Collective Behavior.* 3rd ed. Englewood Cliffs, NJ: Prentice-Hall.

Velotti, Lucia, Tricia Wachtendorf, Natacha Thomas, and José Holguin-Veras. 2011. "The Impact of Security Concerns on the Distribution of Relief Following the 2010 Haiti Earthquake." Paper presented to the International Research Committee on Disasters Meeting, Broomfield, Colorado, July 13.

Wachtendorf, Tricia. 2004. "Improvising 9/11: Organizational Improvisation Following the World Trade Center Disaster." PhD diss., University of Delaware, Newark.

Wachtendorf, Tricia, Bethany Brown, and José Holguin-Veras, 2013. "Catastrophe Characteristics and Their Impact on Critical Supply Chains: Problematizing Materiel Convergence and Management following Hurricane Katrina." *Journal of Homeland Security and Emergency Management* 10 (2): 497–520.

Waugh, William L., and Richard T. Sylves. 2002. "Organizing the War on Terrorism." *Public Administration Review* 62 (S1): 145–153.

Weather Underground. 2015. "Weather History for KNYC, Tuesday, September 11, 2001." Available at http://www.wunderground.com/history/airport/KNYC/2001/9/11/DailyHistory.html?req_city=New+York&req_state=NY&req_statename=New+York&reqdb.zip=10001&reqdb.magic=5&reqdb.wmo=99999. Accessed May 14, 2015.

Weick, Karl E., 1987. "Organizational Culture as a Source of High Reliability." *California Management Review* 29 (2): 112–127.

———. 1990. "The Vulnerable System: An Analysis of the Tenerife Air Disaster." *Journal of Management* 16 (3): 571–593.

———. 1993. "The Collapse of Sensemaking in Organizations: The Mann Gulch Disaster." *Administrative Science Quarterly* 38:628–652.

———. 1995. *Sensemaking in Organizations*. Thousand Oaks, CA: Sage.

Weick, Karl E., Kathleen M. Sutcliffe, and David Obstfeld. 1999. "Organizing for High Reliability: Processes of Collective Mindfulness." *Research in Organizational Behavior* 21:81–124.

———. 2005. "Organizing and the Process of Sensemaking." *Organization Science* 16 (4): 409–421.

Wildavsky, Aaron B. 1991. *Searching for Safety*. New Brunswick, NJ: Transaction Books.

Wood, Stephen. 1987. "The Deskilling Debate, New Technology and Work Organization." *Acta Sociologica* 30 (1): 3–24.

Index

James Kendra is a Professor in the School of Public Policy and Administration and **Tricia Wachtendorf** is an Associate Professor in the Department of Sociology and Criminal Justice at the University of Delaware. They are the Directors of the Disaster Research Center.